HEIRPOWER! 2.0
Eight Basic Habits
Of
Exceptionally Powerful
Lieutenants

BOB VÁSQUEZ, CMSGT (RET), USAF

W A J BOOK PRESS

CONTENTS

FOREWORD

It's November of 2005. I'd written what you're about to read, literally, on a napkin. At least the outline. (I've read that some brilliant writers have done that.) I'd produced a draft copy that I was editing and carried it with me wherever I went, just in case I got another idea to share with you. I'm on a flight from Dulles International Airport to Buckley Air Force Base with former Air Force Chief of Staff and friend, General Ron Fogelman. Just he and I. Oh, and Colonel Jerry Limoge. There's a story there that you can read in my book, *Beyond the Little Blue Book*, which you should read.

I show General Fogelman the draft of my book. He looks it over. Evidently, likes what he sees, and suggests I send it to Lieutenant General Steve Lorenz, who was the Commander of Air University at the time. I did just that when I got back home. Time passes....

It's now January of 2006. I return to my job as a Program Director at the Air Force Academy's Center for Character Development (It eventually became the Center for Character and Leadership Development.) after taking a couple of weeks of leave. As I enter my office I notice the little red light on my office phone blinking, which means that there are messages awaiting my attention. My message box is full. I start listening to them, first one first. It's General Steve Lorenz, the commander of Air University, former boss, and longtime friend. The message is short and sweet, typical Lorenz style... "Chief! Lorenz here! We're publishing your book!" CLICK! That was it. I'm kind of excited but at the same time bumfoozled as to exactly what the general was telling me. Air University owns AU Press. And they publish books. So, I figure that what the general is telling me is that AU Press will be publishing my book. My FIRST book, by the way. (At this writing, I've published twelve books. They're on amazon.com.)

I've been around a long time. I served almost 31 years on active duty and served for many senior officers, including generals, including General Lorenz. The next maybe-50 messages on my phone were from all kinds of colonels reiterating what General Lorenz had said, and I understood. AU Press would be publishing my book. Each messenger suggested I call them back. When I talked with them, each colonel, forwarded me to another person who would begin the process. After a couple of days of doing that, I finally talked with an editor who had been assigned to make it happen. Dr. Marvin Bassett was a pleasure to work with. I had no idea what it took to publish a book, but Dr Bassett did, and he led me through the process in a very professional and respectful way.

That's how the first version of HEIRPOWER! was born. In the years since, I've continued to think about what I first wrote about and how those ideas have continued to grow and evolve. In some ways. The original eight habits are foundational. They're still viable. And if you aspire to be an Exceptionally Powerful Lieutenant, focus on those and I guarantee you that you'll excel. There are, surely, more habits that you should develop but these eight will get you started on the right path. So, what I've done in this version is expand on what I wrote in the first and have added more thoughts, mine and several other Effective Leaders' who have provided their thoughts directly or indirectly. I also, by the way, created a podcast on Apple Podcasts. It's called HEIRPOWER! 2.0. They're interviews with boots-on-the-ground leaders who are out there protecting our country daily and who have read the original book and vouch for its tenets.

I'm including the endorsements that General Fogelman and Chief Master Sergeant of the Air Force, Bob Gaylor, wrote for the original version since they're still my very respected and valued friends and fellow warriors.

7

Thank you for reading my thoughts. Thank you for what you do every day to keep America free. And thank you for who you are.

HEIRPOWER! ALL IN!

bob vásquez, CMSgt (Ret), USAF
Monument Colorado
November 2022

Chief Bob Vásquez has found an innovative and effective way to share some basic principles that every new lieutenant should know on the subject of how to succeed as a leader in our great Air Force. He provides the enlisted perspective in a way that only a senior noncommissioned officer can communicate. I've known the Chief for many years and have seen him succeed as an enlisted leader and as a mentor to the many officers he's served. As a member of the Air Force Academy's Center for Character Development, he mentors our future leaders on a daily basis. It is obvious that serving is his passion. I'm convinced that any lieutenant who reads this book will be better prepared to lead at every level.

RONALD R. FOGLEMAN, General, USAF, Retired
Durango, Colorado
9 March 2006

The odds are that most of us who have served lengthy careers as senior noncommissioned officers in the Air Force have sage advice to share with fledgling officers. The difference is that Chief Bob Vásquez has done it—in a well-written, common sense how-to book. Easy to read and appropriately interspersed with humorous guidance, *Heirpower!* is not just for new lieutenants, but for anyone in a position of leadership. I've known Bob for over 30 years. Not only does he talk this stuff, he practices it daily—and it works. After you read *Heirpower!* do what he says, and you'll be an *exceptional* leader.

ROBERT D. GAYLOR, Chief Master Sergeant of the Air Force, Retired
San Antonio, Texas
18 January 2006

People may be impressed by what you do, but they're inspired by who you are!

bob vásquez

BOB VÁSQUEZ

ABOUT THE AUTHOR

Chief Master Sergeant, Retired, Bob Vásquez is a 50-year veteran of the United States Air Force. He served almost 31 years on Active Duty and more than 19 as a program manager, instructor, and curriculum developer at the United States Air Force Academy. He served as the course director for the freshman seminar offered through the Academy's Center for Character and Leadership Development. He retired from Active Duty on 31 October 2002 and from the Air Force Academy on 28 February 2022. Bob invested 24 of his active-duty years serving with Air Force bands throughout the world. He also served as Commandant of the Noncommissioned Officer Academy at March Air Force Base, California; Senior Enlisted Advisor to the Commander of the 92nd Air Refueling Wing at Fairchild Air Force Base, Washington; Deputy Director of the Family Support Center at Ramstein Air Base, Germany; and Superintendent for the 86th Support Group at Ramstein. Bob is also an adjunct instructor at the University of Colorado at Colorado Springs.

He's the author of *So Ya Wanna Be THE Chief?!*, *S.S.G.T.*, *The College Freshman's Beyond Survival Guide!*, *What I Learned from Dad Made Me a Better Man!*, *The Power Of SUPERvision! Beyond the Little Blue Book, A Different Shade of Blue, and others.*

A wisdom seeker, igniter, storyteller, musician, public speaker, life coach, and mentor, Bob considers his greatest accomplishments the successful raising of his daughters, Tesa and Elyse, six grandchildren, two sons-in-law, and growing closer to Debbie, his lovely bride of more than 45 wonderful and fulfilling years.

Connect with bob at bobvasquez@bobvasquez.com.

PREFACE

Isn't this a great day to be an American Warrior?! If I had a dime for every time I've said that, I'd be driving a new JEEP! Wait a minute! I have *three* new Jeeps in my garage! I've gained some notoriety from that question, but that's not why I ask it—nor why I developed it. It's a sincere question of loyalty and dedication to a way of life that I enjoyed for most of my adult days. I've written this book to share my experiences and joy with you and to help you be your best because I believe that if you're an effective officer/leader (aka an Exceptionally Powerful Lieutenant), you will, certainly, take care of a group of people I love dearly: enlisted folks.

I'm not sure you got the enlisted perspective from the institution where you received your commission, but you're going to get it now! In a good way, of course. My goal is to help you empower yourself with habits that will make you an Exceptionally Powerful Lieutenant. What I'll share with you comes from an enlisted perspective. Don't throw away what you already know about leading, just add this material to your toolbox.

But enough about you. Let's talk about me. I served in the United States Air Force for more than 50 great years, 30-plus on active duty, and 19-plus as a civil servant at the United States Air Force Academy. The official line is that I retired from active duty in 2002, but the term *retirement* carries a tinge of volunteerism. I did not volunteer to leave my service. I would have stayed another 30 years had I been afforded that opportunity. I love serving with people whose purpose lies beyond their own desires and who do everything in their power to keep our country free.

I served in the rank of Chief Master Sergeant (E-9) for almost 12 years—more than a third of my service time—and, as such, had many opportunities to mentor officers of every grade and every rank of enlisted person. I can't tell you how many lieutenants I chewed up and spit out! (Okay, I never did that, but it sounded good and got your attention, right? Besides, I can't tell you because I can't remember. It'll happen to you one day. Give it 60 years.) Speaking of which, I still always stand at attention for lieutenants, who have a special place in my heart because they're the first officers who lead my enlisted brethren. I'm convinced that that first leadership opportunity sets the stage for the rest of an officer's military life, eventually leading him or her to become either a good, or bad, senior officer. I hope that this book will help you become a good senior officer.

I invested more than 19 years observing people like you, young future leaders, as a member of the Air Force Academy's Center for Character and Leadership Development, creating and leading workshops, and as a member of the Commandant of Cadets' Curriculum Division, as well as just hanging out inspiring the next generation of leaders, mostly by observing and listening to what Cadets were concerned with and interested in. I loved every moment I got to be with those Leaders of Character. "How did you ever get there?" you're asking. Here's what happened....

I was in my last year on active duty. Hitting the 30-year point, I knew I had to retire. I'd submitted a request for an extension, but it was denied. I had to get out. I didn't know what I would be doing next and since my family and I were stationed at Ramstein Air Base in Germany at the time, I had no network to fall back on. Then 9/11 happened. What happened that day changed the world in many respects for all of us. I was supposed to retire at the end of January 2002. I'd entered the Air Force on 1 February 1972. I HAD to retire at the end of January. But WAIT! Not so fast! All of a sudden, the Air Force implemented STOP LOSS, which meant I COULDN'T retire! No one could. That was okay by me! So, I ended up serving an additional nine months beyond the 30 I was supposed to. That's not the entire story. Here's the rest of it....

On September 11th of 2002, a year after 9/11, a few hundred of us are on our way out to the flagpole in front of Wing Headquarters at Ramstein, en route to the Retreat Ceremony commemorating 9/11. Since my office at the time was across the street, I was walking to the flagpole when Colonel Paul Valovcin, the Logistics Group Commander, approached me and asked me what I'd be doing once I retired. I knew a lot of folks at the base and pretty much everyone knew I was supposed to retire. As the good colonel asked me, I replied that I had no idea. I'd submitted a few resumes here and there but hadn't had a bite yet. Remember that this was in 2002. Our technologies weren't anywhere near what they are today. Colonel V was a graduate of the United States Air Force Academy. He told me that the Academy had an opening in its character development cadre. He suggested I apply. "Sir," I told him, "I'm an enlisted guy. They wouldn't want me." "Will you apply for ME, Chief?" he asked. "Of course, I will, Sir...for you!"

I had to figure out how to get onto the website to apply for the position, which I did. I applied and eventually got an interview with Lieutenant Colonel Russ Sojourner, the Chief of the Cadet Development Division. I was one of 60-plus applicants for the job. But I got it! I suppose that it helped to have some support from people like General Steve Lorenz,

USAF, Retired, and General "Speedy" Martin, USAF, Retired. I'm told that General Lorenz called the Academy and told the deputy of the Center for Character Development, Colonel Tom Berry, USAF, Retired, to "hire him or else!" That sounds like General Lorenz. A great officer with whom I served for many years.

I was charged with teaching young men and women whom we called Cadets to become Leaders of Character. I learned as much from them as they might have learned from me. And I had many opportunities to go out to Air Force installations and learn about other commissioning sources and how they developed leaders of character. We all had different methods, but the goal and the purpose were the same.

What gives me the right to make the assertions I do in this book? As I considered my answer to that question, I tried to remember when I started leading people. It was almost easier to remember when I *didn't* lead. I've led for as long as my memory will take me back. That, in itself, doesn't make me an authority, I know. I've also observed and studied countless leaders, taking note of what worked, and what didn't work, for them. I was blessed with working with both the best and the worst. I learned a great deal from both types.

You may not agree with all that you read here, and you're welcome to argue with me if you please. You may think that "here's an Old School Chief telling us young officers how it was done back in the day." You may already know all that I tell you. You just need a reminder. Excellence in anything you endeavor will always be based on the fundamentals. Whether it's in sports, in the Profession of Arms, in developing your personal character, or in leading yourself and others, if you don't know the fundamentals, you'll fail. I can assure you that what you're about to read is the fundamental truth to becoming an Exceptional Leader. My hope is that you'll laugh, cry, and learn as I share some basic leadership ideas with you that will make you an Exceptionally Powerful Lieutenant. Now, go forward! Make this a great day to be an American Warrior.

HEIRPOWER! ALL IN!

bob vásquez!
BOB VÁSQUEZ, CMSgt (Retired), USAF
Monument, Colorado
24 November 2022

ACKNOWLEDGMENTS

Everything we do in this world is touched by someone else. I'm grateful to everyone who has nurtured me and given me opportunities to excel and to fail. I learned from every experience. The following is only a partial list of those to whom I owe much. There are many more, but it would take another book just to mention them all.

Thanks to Dr Russ Sojourner, Maj Jeff Kozyra, Col Tom Berry, Maj Gen Gary Dylewski, and my colleagues at the US Air Force Academy's Center for Character and Leadership Development who continuously supported me in making my vision a reality.

Thanks to my personal life coaches who continuously show me the way: Mr. D. J. "Eagle Bear" Vanas, Mr. Noah benShea, and Ms. Sarah Hurd.

Thanks to the officers who have mentored and inspired me through the years: Gen Ron Fogleman, Gen Stephen Lorenz, Maj Gen Gary Voellger, Col Denny Layendecker, Col Eric Carrano, Col Paul Valovcin, Lt Col Doug Monroe, Lt Col Hank Emerson, and all of the lieutenants, too many to list, with whom I stood to make a positive difference.

Thanks to the Chiefs who made me a Chief: Chief Master Sergeant of the Air Force (CMSAF) #5 Bob Gaylor, CMSAF #11 Dave Campanale, CMSAF

#1 Paul Airey, CMSAF #7 Bud Andrews, CMSAF #8 Sam Parish, CMSAF #6 Jim McCoy, CMSAF #9 Jim Binnicker, CMSgt John Sterle, CMSgt Bob Smith, CMSgt Roy Boudreau, CMSgt Ron White, CMSgt Frank Guidas, CMSgt Mac McVicar, CMSgt George Moriarty, and CMSgt José Tavarez.

Thanks to all of the cadets who have passed through the US Air Force Academy. Those I've known are the real inspiration behind this book. They are the best and brightest who endure much more than I ever could so that they can fulfill their dreams of being warriors and defenders of our way of life and freedom.

Thanks to Mom, Dad, Tío Rulie, Momma Joyce, and my siblings for providing me the love and nurturing I needed to succeed as a warrior. Thanks to the two teachers who empowered me to grow—Mr Bruce Firkins and Mr Jack Hall—and to my editor at Air University Press, Dr. Marvin Bassett, who helped make sense of what I've written.

Thanks to Maj Henry Sims, Maj Lisa McCarthy, CMSgt Joe Bogdan, SMSgt Mealinda Koory, MSgt Josh White, and TSgt Dalven Adams for providing their War Stories.

A BIG thanks to my First Protégé and publisher, CMSgt (Ret) Dr José Lugo Santiago, and My People, Wendy Lugo Santiago, from WAJ Book Press.

Special thanks to Debbie, my lovely bride of 45 wonderful and fulfilling years, who keeps me on the right track; who endured all of the assignments,

TDYs, and late-night calls; and who gave me the greatest gift of all: my daughters, Tesa and Elyse. Thanks to my grandkids Nieves, Alexia, McKinley, Marisa, Malayna, and James, and their dads, Jeremy and Nathan.

Lastly and mostly, I thank the Creator at least twice a day for the blessings I receive every day. I'll never be worthy, but I accept them as Grace.

BOB VÁSQUEZ

INTRODUCTION

You may be wondering how I chose the title for this book: Heirpower 2.0! Eight Basic Habits of Exceptionally Powerful Lieutenants. Or you may not. Here's the explanation anyway. I'll break it down one word at a time because I believe words can be powerful. It may help if you understand the words in the title because they bind the text into my purpose for sharing these thoughts with you—to help you become an Exceptionally Powerful Lieutenant.

A lieutenant is a leader—or is expected to be one—so please forgive me if I use those terms interchangeably. By the way, although the title may sound similar to others you may have read, the content is not. Believe me.

The concept of heirpower is simple. An *heir* is a *successor*. If you break that word apart, you're left with *success* and *or*. *Success or*.... The alternative to success is failure. You ever wake up in the morning and your first thought is, "Hmmmm.... What can I really screw up today?" NO! You don't do that! That's a failure mentality. This book is about success, not failure! What you'll find here deals with success—not failure. Better yet, it's about enabling people to empower themselves to succeed.

*Powe*r is the capacity to act effectively. Who has power? You do! I do! We all do! **Real power is produced when we share what we have.** I'm going to share all that I can with you that will enable you to succeed. You'll achieve real success when you take what I give you and what you develop on your own and then share it with others. The only way to get is to give. **The best way to succeed is to help others succeed.** That's what an effective leader does. Pass on all you can, and you, too, will develop heirpower.

Why the number eight? My daughters used to watch a television program called Schoolhouse Rock. They loved that show. Okay, I loved that show, so they had to watch it. One of the show's songs—about the figure eight caught my attention. One of the lyrics went, "Figure eight as double four . . . that's a circle that turns 'round upon itself." What's that got to do with leadership? Well, CMSgt Gene Gardner, USAF, deceased, an old friend of mine, once told me that **"as leaders, we can do one of two things: we can do something to our people, or we can do something for our people."** Now, as we get to know each other through this book, you'll find that I enjoy playing with words, so let me take his statement and make it more powerful. Isn't for (four) twice as much as to (two)? Hence, eight will be even twice as powerful as four (for), right? If that doesn't make sense, don't worry about it. I'll probably tell you more things that don't make sense on the surface. You'll just have to trust that I know what I'm trying to tell you. You'll also find that I like circles. The figure eight—a circle that turns 'round upon itself—is powerful too. You'll find at the end of this book that these concepts all turn 'round upon themselves. They're simple but important, and they all almost blend into each other. **Leadership is a cycle that continues indefinitely.** I'm convinced that by applying these eight important leadership lessons, you will become an exceptional leader, especially if you make them habits.

Before I go on, let me share my thoughts on leadership and leading. Leadership and leading are two different things. Leadership is a science. Leading is an art. I'll, surely, use both terms interchangeably because you'll need both to succeed, and there's value in both. You really can't succeed without both. The real value, though, is in you applying these ideas by leading yourself and your followers. I'll provide you the science, if you will, by sharing thoughts, and ideas, but you'll have to practice the art of leading, assuming you intend to be an effective leader. The ONLY way to become an effective leader is by leading effectively. You have to do the work.

The term *basic* is, well, basic. I studied with a professor who taught me that "life is simple, once you understand its complexities." I've told you that my purpose is to help you become an Exceptionally Powerful Lieutenant. An important question is, "Who will you lead?" Yourself, and other people! Pretty simple, don't you think? Maybe not. People are complex entities. You're a complex being, aren't you? We even take pride in that sometimes. Leading, really, is much simpler than we often think. The problem is that we don't think—neither very much nor very often. When we do, it's not always about what we should be thinking about. As you read this book's eight admonishments, you may say, "Hey, I knew that!" That's okay. I don't profess to be a prophet. I'm just trying to guide you to think about what will help you become an Exceptionally Powerful Lieutenant. As Samuel Johnson once said, "People need to be reminded more often than they need to be instructed." You probably already know this stuff. You just haven't thought of it lately. See how basic that is? Or is it too complex?

I believe Aristotle said, "We are what we repeatedly do." Dr. Stephen Covey, author of *The Seven Habits of Highly Effective People*

(Yeah, it sounds like the title of this book, but don't forget I told you the content is different—so read on), said that we first build our habits, and then our habits build us. There's a lot of truth to what those sages said. The more we do something, the better we become at doing it. That TV ad that says "Just do it!" doesn't necessarily have any thought behind it. I'm hoping that this book will first make you think about why you should do it before you do. And then you can do it right and you can do it now! I'll even tell you what *it* is. Habits, like words, are very powerful in that, eventually, you won't even think about what you're doing. You'll make things happen almost naturally—habitually. Simply.

In the paragraph that follows, I'll explain what powerful means, but before I get to that, let me ask you…. Are you the type of person who just wants to make the grade? Are you comfortable simply meeting the standard? Or are you willing to do what it takes to achieve your personal best in all you do? Do you see yourself setting the standard for excellence? If you answered *yes* to the first two questions, then go read a Marvel comic book. If you answered yes to the second two, then read on. You're going to be a great leader! An exceptional leader! You see, to lead the enlisted folks we recruit these days, you're going to have to be way above ordinary.

Just meeting the standard will not be good enough. Your followers won't follow you unless they believe you're better than they are. The eight habits I'm about to share with you will set you above everyone else. By living them, you will be—I guarantee—not only powerful, but exceptionally powerful.

We use the term *powerful* fairly regularly. Do you know what it means? I'm going to bet that you've never even thought about what it means. Let me help you out. Powerful means full of power. Basically, that's what it means. One dictionary defines *full* as "containing all that is normal or possible." Power, it says, is "the ability or capacity to perform or act effectively." According to Covey, "effectiveness is getting what you want again and again and again." You are full of power! You have within you all of the capacity you need to inspire your followers to follow you. We'll dismiss the word *normal* in the definition of *full* because you're not willing to settle for *normal*. You'll set a higher standard for those you lead because you have the desire and the will to do it. ***This book's eight lessons are basic to leading powerfully. If you do them and make them habits, you will get what you want—the respect and loyalty of your followers—over and over again.*** Even when you're not there! Know this: power is not control. The only person you may control is you, and good luck with that! Real power lies in knowing what the people you lead need and want and providing it to them so that you can all grow together. The lessons from this book will empower you to do that. Let me show you an example of power. Clench your fists in front of you. Go ahead. Put the book down but keep it open so that you can do what I'm asking you to do. Again, clench your fists. THIS is what many "leaders" think is power. It's not! What can you do with clenched fists? Hit something or someone, I suppose. But in attempting to lead do you think that'll work? It almost used to, but not so much anymore. Yeah, here comes a story. But before I share, it let me finish this example. Clenched fists is NOT the way you lead people. Now, unclench your fists and hold your arms out as if you're going to embrace something. Palms open. How much might you hold and affect in that position. As much as you can! THIS is how you lead effectively. You embrace others, at least figuratively. You accept and even celebrate who they are, not who YOU are, and help them empower themselves to be them. What these eight habits will do is empower YOU to guide them toward accomplishing the mission. Again, you'll have to trust me. Here's the story I promised you....

Back in the day, before you were born...waaaay before you were born, in the 1970s, we were often "trained" through fear. I remember my first assignment with the 15th Air Force Band at March Air Force Base in California. Within days of arriving, the three of us Young Airmen, at that time, Mike Mormon, Mike Paulson, and I, were taken outside the Band building by Staff Sergeant Rick Rose, the Duty Sergeant, for him to tell us what our additional duties were going to be. "Airman Mormon, your job is to...." I don't recall, but it was much easier than mine, I assure you. "Airman Paulson, YOUR job is to...." Whatever. Then, "Airman Vásquez, your job is to keep this sidewalk clean." There was a huge eucalyptus tree right next to the Band building. The building and the tree were separated by a sidewalk, which was now my responsibility to keep clean. That tree shed its bark like it was going out of style. That sidewalk was constantly strewn with bark. I noticed it the first time I walked up to the building. Sergeant Rose gave me specific instructions. "Here, Vásquez, take this broom and sweep that sidewalk. I don't want to see any bark on it. Ever." "WHAT?" I thought. Evidently, I wasn't just thinking it, I was saying it, and more. "WHAT? You want me to sweep that sidewalk every day and keep it clean every day, all day long? Are you crazy (I thought I was just thinking.)?" In those days, it was a big thing to have gone to college prior to enlisting. You graduated from Basic Training as a two-striper, an Airman First Class, vice an Airman Basic no-striper. I'd gone to New Mexico State

University for three semesters, which made me eligible for the extra stripes. "Look, Sergeant Rose," I told him, evidently, "I'm an Airman First Class. I went to college. I'm a musician. I'm a bassoonist. I don't do that!" Again, I was thinking I was thinking, not saying. Man, did that set Sergeant Rose off! I can still feel it...he sticks his finger in my chest and tells me, "Do it or I'll kill you!" Let me just say that that sidewalk was the cleanest sidewalk on base for as long as I was there. The lesson here is that, back in the day, that's how we were trained. That doesn't work anymore. You can't even think those thoughts. But that's okay. As we evolve, we have to devise different, hopefully better, ways of leading people. Keep evolving....

If you develop these eight habits, I assure you that you will succeed as a leader, particularly a leader of enlisted people. The enlisted men and women you lead will follow you and nurture you. If you haven't heard this next statement already, you will soon: "The enlisted corps is the backbone of our military forces." Enlisted folks have a different function than officers, but they have the same value. Leaders must realize that they cannot do everything and, in truth, can do nothing without the support of their enlisted people. Lieutenants lead enlisted people—not officers. An effective lieutenant does the eight things I'm about to share with you. Read this book. Learn what it says. Do it on a daily basis. And enjoy the ride! I'll try to emphasize this in Habit 8. Find an enlisted person or two, preferably a Senior NCO, whom you can trust to guide you and train you in these eight concepts. Help them help you by honestly assessing whether or not you're living up to the lessons and how to improve when you're not.

If what you've just read sounds familiar, thanks for reading the original version of this book, *HEIRPOWER! Eight Basic Habits of Exceptionally Powerful Lieutenants*. It was published in 2006 by Air University Press. This version includes the same foundational material that the original book included only with an updated perspective and some additional thoughts and a lot more stories. As I said earlier, I've produced a podcast that I called HEIRPOWER! 2.0 (amazing, ha?) that's available on Apple Podcasts. It's a series of interviews with boots-on-the-ground Senior NCOs and officers sharing their perspectives on the original habits (which, by the way, haven't changed). They've each given me permission to share some of their thoughts and even share some War Stories to update the tenets of the original book. I commend you to listen to the podcast for additional guidance.

Oh, you're thinking I left out an explanation of the last word of the title—lieutenants. Believe it or not, that was intentional. Up until the day you walk into your first—or new assignment, everyone will have a vision of what a lieutenant is. That vision is not necessarily positive. My hope is that having read this book and having made a commitment to follow its tenets, you will redefine what people think a lieutenant is—and can be. You will be so good that when you leave your unit, people will say, "That' MY Lieutenant! And s/he's one Exceptionally Powerful Lieutenant!"

BOB VÁSQUEZ

So, you're 22 years old, you've just gotten your commission, you've arrived at your first duty station, you've met with your commander, and you're now "in charge" of a group of enlisted men and women, all of whom have been in service longer than you, know a whole lot more about military life than you, and are expecting more than you know. To top it all off, your first "subordinate" happens to be a 30-year veteran of every war you ever read about, and his rank is, yes, E-9. He's not an E-10 only because that rank doesn't exist. Now, what do you do? Let me tell you....

HABIT 1

GET A HAIRCUT!
FIRST IMPRESSIONS LAST

Yeah, things have changed. Some things. How we make a first impression has evolved. Yeah, that's a good word. The impact of a first impression has not. It's still very powerful.

Back in the day, and it's still true to a large extent, we made a first impression in person, physically. Nowadays, we often make a first impression digitally online, as well. How ever you do it, first impressions last! And as my mentor, Chief Master Sergeant of the Air Force #5, Bob Gaylor, taught me, high tech will never replace high touch. It won't. But you already know that. As a leader, you'll have to develop and practice that skill. The high touch one. Maybe both.

Standards have changed and evolved, but as the leader, you have to not only set the standard; you have to surpass it, you have to model it, you have to BE that higher standard. Think about this. When we think of standards we think of MINIMUM requirements. Minimum anything always creates mediocrity. Is that your goal, especially as a leader, to be mediocre and help your followers to be mediocre?

I hope not. Probably not… Okay, I'll answer for you…HECK NO! Setting that standard as a leader begins with the first impression.

"Great news, Warriors!" the commander says excitedly, "I brought you all here today to introduce you to your new leader. You all know we've been looking forward to this for a long time. Please do all you can to make the lieutenant comfortable."

This is the day you've been waiting for. After months—even years—of preparation, you've finally made it! You are the leader! The commander has pumped you up. You've pumped yourself up. You're anticipating that your followers will be as excited to see you as the commander obviously is. You. Are. Ready!

Oh, WAIT! As you climb the steps to the lectern, you get a glimpse of the crowd gathered to greet you, and you see something different than you expected. They're not happy. What can they be thinking? "Aw, man!" the enlisted folks are thinking (actually, they're saying it), "Another green lieutenant. Here we go again!"

As you reach the lectern, you look out at the followers and see a change of face. You might as well be wearing a superhero outfit because you are tight (that's Old School for sharp). You look good. Your hair is perfectly coifed and tapered beyond standards. Your uniform is so crisp that it snaps, crackles, and pops (yes, I'm making reference to Rice Krispies). Your shoes are so shiny they compete with the glare from the Old Man's bald head. I mean, as you walk onto the stage, the theme music from *Rocky* starts blaring out of the intercom speakers. Okay, wake up!

I was about to say that I hope your first duty day is like what I just described. The truth is that you can *make* it that way. It takes work, though. As I tried to emphasize in describing the word *habits* in this book's title, the idea is to do what's right so often that it becomes habitual. You don't even have to think about it—you just do it. Interestingly, that phrase is closely related to Nike products. You may not know it, but Nike was the Greek goddess of victory. (Okay, so you did know that. Hey, I'm trying to impress you here!) Victory in life comes from developing good habits. Being good often begins with looking good, and your followers will expect you to look at least as good as they do. Now think a little bit. As the leader, don't you think you should look even better than the people you lead? Hmmm....

Whatever first impression you make on your followers, whatever they think you look like the first time they meet you, will stay in their minds forever. For. Ever! First impressions last! One of the reasons we insist that our military members wear their uniforms properly, wear their hats outdoors, and so forth, is that it presents a professional military image. You are a professional, aren't you? You are in the military, at least for the next several years, right? Your personal appearance will set the standard for your unit. What you give, you'll get. If you look good, chances are your followers will emulate you. If you look sloppy, chances are your followers will emulate you. That's the truth. So, **make it a habit to look your best, not just good, every day.** Be exceptional!

You'll also make an impression by the language you use. The military doesn't condone the use of vulgar language. Guess who has to enforce that? You! You can't enforce that rule if you break it yourself. Vulgar language validates a person's ignorance and disrespect for others. The American language includes myriad terms that you can use to provide the emphasis that vulgar words do. People's use of offensive language becomes habitual. Most of the time, they don't even realize they're offending anyone—but they are. Even if the offended ones don't say so. Live up to the standard of Effective Leaders and get rid of vulgarities in your speech. Think about this: where do vulgar words come from? Vulgar minds. Get rid of vulgar thoughts, and you'll get rid of vulgar words. Don't use them! Remember, you are now a leader—a professional leader. Think like one and speak like one.

Your followers will be looking out for if—and how—you respect them. If you want to start out on the wrong foot, call them all by their first names. If you want to impress them, call them each by their ranks and last names. Remember, you're setting the standard. What may happen, depending on the professionalism of the group, is that they'll ask you to call them by their first names. You decide. Is that a professional habit to develop? It may depend on the maturity and trust levels within the unit. Nonetheless, don't take it upon yourself to call your followers by their first names. If they give you permission to do so, consider it deeply before you do it. And *never* give a follower permission to call *you* by your first name! I've often heard people say that they have different rules when they're "off duty." "It's okay to call each other by first name when we're having a drink," they say. More on that later. First, you're NEVER off duty. Most importantly, if you try to be different at work and at play, you'll eventually mess up. I've been there way too often. You'll do what you practice. Practice being respectful.

I'm a musician by trade. I was a Band Guy for 24 of the 31 years I served on Active Duty. I learned the musician vernacular and used it daily. To the point that I used the F bomb pretty much every other word. It was the musician vernacular. I always thought, even said, that I could control that. I knew that I would never use vulgar language around my loved one, my family. Until one day….

I was at my Tío Rulie's home, visiting him and my aunt, Mama Joyce, when he asked if I'd like to join him as he walked his dog. I accepted his invitation. Tío Rulie was my favorite uncle in the whole wide world. He was like a second father to me. I spent my high school summers with him and Mama Joyce. I loved them both dearly. As we were taking our stroll, I started telling Tío a story of some sort. I guess I got a bit excited because, all of a sudden, without realizing it, I used the F Bomb in the midst of my conversation. Now, that wasn't unusual in those days, except that I would, as I just said, NEVER have even considered saying it in front of my Tío Rulie. I said what most people say, "I won't use it when…." Well, it didn't work. I DID use it when…, plain as day. And I immediately saw the effect it had on my beloved uncle. He looked at me surprised and disappointed. I'll never forget that look. We never discussed it. I moved on as soon as I could so as not to bring any more attention to it. It was done. Nothing I could do about it now. But I never forgot it. I'm sure he didn't, either. I disrespected my uncle unintentionally.

We are what, and who, we are whether we're at home, at work, anywhere. If you develop and practice bad habits, especially in your language, you're going to fail sometime. You're better than that. You know the difference. By the way, I promised myself at that moment that I would never use profanity again. I haven't to this day. Sometimes, it's very difficult, like when I've stubbed my little toe on the couch, but so far, so good.

You'll also impress your followers, at least the professional ones, by the way you salute. I'm amazed at the different types of salutes I see officers render. Okay, often, enlisted folks aren't much better. I'll admonish them in another book. Your salute sends a clear message about how much pride you take in your profession. The military profession is the only one in which people salute each other regularly. Your salute says a lot about how you view discipline. I'll tell you more about discipline in a later chapter. A shabby salute says you don't care. That's the truth. If you salute proudly, sharply, and appropriately, everyone will see that you mean business when it comes to military bearing. Oh, one more thing on saluting: don't thank an enlisted person for saluting you. You'll never find an enlisted person thanking an officer for returning a salute. It's a sign of mutual respect. Leave it at that.

Is it possible to make a bad first impression but repair it? It is, but it will take a lot of work and a lot of time. You can repent and do all sorts of good things for your folks, and they may show you respect. But know this: if you ever do anything dumb, they will revert to their first impression of you. "I always knew the lieutenant was a dirtbag. I remember the first time I saw her. I knew she was bad!" That's what will go through their minds and their lips. Trust me! I've been there. First impressions last!

As I said at the beginning of this habit, things have changed. More and more, we make first impressions online and on our smart phones.

I know there's a way and reason unknown people gain access to my email address. I wish that would stop, but it doesn't look like it will any time soon. It's amazing how many uncles and ambassadors from other countries email me daily to tell me that there's a multi-million-dollar reward awaiting me and that they will deliver it if I send them money or information, or both. I won't get into specifics on how to make a good first impression online because you probably know how better than I. Or how to make a bad first impression.

Courtesy and professionalism are key. ALWAYS proofread your emails before you send them and ALWAYS, ALWAYS, make sure you're sending them to the right people. Don't Reply All unless you're absolutely sure it's what you intend to do. I remember Replying All once with an inside joke that the person I thought I was replying to would understand. Well, the other hundred-plus folks who received that note didn't find it amusing. I was pretty embarrassed. There are plenty of books online that can help you write emails in a professional way. Get them. Study them. And practice what they teach. It'll pay off one day.

If your smart phone is the method by which you make a first impression, note that what the listener hears is what they believe. As I already said, the language you use will be noticed and noted. Be professional. Your attitude, too, will shine through on that phone call, whether it's a conversation or a voicemail message. I don't know about you, but there's a very small chance I'll return a call to someone who sounded rude on the voicemail they left me. And maybe it's just me, but I listen to the background of a voicemail as well. If I hear a bunch of noise, especially noise that seems disrespectful in some way, I'll delete it and never reply.

And then there's social media! BE CAREFUL WHAT YOU POST! It becomes evidence! We all want to practice our First Amendment right of freedom of speech, but as Dr Stephen Covey taught me, "The truly free are the disciplined." If you choose to share your opinions, (which may or may not be valid, but it doesn't matter, you put them out there for everyone to see) they can bite you in the rear end. Once again, I'll remind you that, as the leader, you have to exceed the standard. There are many times that I'm irked by some of the posts I see on Facebook (Yeah, I'm Old School!) or LinkedIn that I so want to respond to. But I remind myself that I have to remain disciplined. I'm seen as a leader by many folks. I have to BE the standard and I wouldn't want to read what I would like to post coming from my protégés. It's tough sometimes, but it's the right thing to do. I'm supposing that social media is probably THE way you Youngsters (Old School term) make a first impression. Before you do, consider how you want to be perceived. Choose wisely.

I'm sitting in my cubicle at the shop at the Air Force Academy, minding someone else's business online, when I hear the voice of one of most favorite Cadets who just graduated a few months ago. Don't judge, but Academy Grads get 60 days of leave right after graduating, which is well-deserved in my opinion. They'll go off to exotic places, like Pueblo, and enjoy some freedom. It was right about the 61st day point when I hear that LT's voice. As I turn around, I see him. He's in ABUs (this was a few uniforms ago), big smile on his face, and a HUGE mustache and sideburns that match his hair. "Hey, Chief! I'm back!" he announces. I give him a man-hug and ask him to take a seat so that we can catch up. Once a Chief, always a Chief. I can't help myself. I want to make sure I'm not applying Old School rules, though, so I call our Senior NCO over to my cubicle. MSgt John Grijalva, one of the most professional Senior NCOs I ever worked with, comes over. He greets the LT as his eyes bulge out! I know what and how he's wanting to address the LT but being that this is a new, kinder, and gentler Air Force, he, thank goodness, softy tells the LT that he's out of regs. He needs a haircut, his mustache is too long, and his sideburns are way out of regs. They resemble Elvis Pressley's. Now you're thinking that that's a good story, but it's only the beginning.

As the LT, Master Sergeant Grijalva, and I are chatting. Someone interrupts us to tell us that the Superintendent, a three-star general, is on her way to see us all. We're to meet up in the conference room ASAP. The LT has no way of fixing his appearance. We all head to the conference room. As the general enters the room guess who's the first person she sees? The LT! She looks him over almost like a drill instructor would and says, "Nice mustache…." She moves on and we have our meeting as the LT beams.

After the meeting the LT comes up to me to continue our conversation. He says, "Did you hear what the general said to me, Chief? She said, 'Nice mustache.'" Master Sergeant Grijalva is right behind me. He interrupts us, telling the LT, not as nicely as before, "That's not what she meant, LT. Go shave and get a haircut NOW." We later explained to the new LT what the general meant and the importance of making a first impression. He understood and never needed further guidance the rest of the time he was with us. Last I heard, he's doing well.

A first impression will go far, whether it's a good one or a bad one. Don't risk being able to amend a bad first impression. As General George S Patton said, "You're always on parade." You're always making a first impression. Make a good one! First impressions last!

I invite you to listen to my podcast, HEIRPOWER! 2.0, on Apple Podcasts, you'll hear a few of my guests mention the importance of humility in making a first impression, especially with Enlisted Warriors. Dictionary.com defines humility as "the quality or condition of being humble; modest opinion or estimate of one's own importance, rank, etc." The question is, "Who knows?" "Modest opinion or estimate of one's own importance" means that we're the only ones who know whether or not we're being humble. Here's some advice from my book titled, *Beyond the Little Blue Book*, on humility.

Maybe we avoid discussing humility because it's very difficult, in fact, near impossible, to assess or measure. Think on that. How would you measure humility? How would you assess that someone is humble? I think you'd say, "by how they do things." But have you ever done something in all humility that someone considered selfish or even arrogant of you?

If we're going to measure humility it can't be on someone else's terms. As Pope Francis said, "Who am I to judge?" I'm convinced that we're the only one who can measure our humility. You know when you've done something in a selfish way or in a humble way. It's the first definition of integrity...doing what's right when no one can judge you.

Look good, speak well, be respectful, salute sharply, and watch what you post online. Start now to develop these habits. Pay attention to whether or not you present a professional military image. Ask someone you trust—not necessarily a buddy—to help you. Make sure you look your best all the time—even off duty—because when your followers see you downtown, they will talk about you back at the shop. Help them say good things about you. If the people you hang with use vulgar language, either make them quit or find new friends. The only way to gain respect is to give it, so treat your followers with respect on and off duty.

Be proud of your chosen profession and express that continuously. If you need to practice your salute, stand in front of a mirror and do it. But do it right! Remember that old saying "Practice makes perfect"? Wrong! Practice makes permanent. If you practice doing something badly, you'll soon be doing it badly—but perfectly badly! I think you get the picture. Be professional in your social media postings. There, like everywhere else, you're the leader.

You'll find two types of enlisted folks in your unit: those who need you to lead them and those who can lead *you*. I'll tell you more about that later. They need you to lead them in the right direction. In the past, I would have admonished you to aim high. What the heck, aim high! Make your standard continue to rise by raising it; your followers will raise it by following you. You are always on parade. You're representing yourself, your team, your service, and your country. Your followers, peers, and leaders are watching you. You have the capacity to excel. Don't just do it. Do it right! Make it last!

In his book, titled *Blink*, which you should read, by the way, Malcolm Gladwell proposes that we make up our mind about people and things in three seconds. Being the overachiever that I am, I do so in one. Gladwell goes on to say that that immediate first impression may or may not be accurate, whatever accurate means, but it's true.

Here's the bottom line. You may have the best of intentions. You may have the best message ever thought of. You may have the solution to all of your followers' woes. But if you don't deliver it right the first time, it will never take effect. As Will Rogers once said, "You never get a second chance to make a first impression." First impressions last!

War Stories

My airmen came to me asking for help. I call them my airmen, but they actually belonged to someone else several layers of leadership away from me. They didn't have a chief to go to, so they came to me. Their noncommissioned officer in charge (NCOIC) and I had this conversation:

NCOIC: "You have to do something, Chief!"

Me: "What's the problem?"

NCOIC: "It's the new lieutenant!"

Me: "What's wrong with him?"

NCOIC: "He's having us go through open-ranks inspections! He comes around the shops with a white glove on, inspecting our tools and our areas. He expects us to wear clean uniforms every day. Since the first day he got here, he's just been a real jerk, Chief!"

Me: "It sounds to me as though he's trying to raise the standard in your organization. Can you honestly say that what he's expecting you to live up to is bad for your unit?"

NCOIC: "No, Sir."

Me: "Maybe the problem is that your folks have become complacent, so now this new lieutenant's vision of your potential is going against that and you're having a hard time dealing with getting back on track." (Man, I was being brilliant!)

NCOIC: "Chief, would you come over and check it out for yourself?"

Me: "Sure."

I did go by. I met the new lieutenant. He looked terrible. Oh, he was wearing a clean uniform, but he surely had slept in it the night before. I know, I know, you're not supposed to starch the Battle Dress Uniform, but you should, at least, iron it before you wear it. His hair was interesting—not exactly a Jheri Curl, but someone must have made plenty of money selling Pomade to this man. It was slick all right, but it, obviously, was holding in the hair so that it didn't look as long as it was. And his language! First of all, he befriended me immediately. Now, I don't want to give you the impression that I'm a snob or that

Chiefs can't be friendly, but when I introduced myself as "Bob Vásquez," he took it seriously! He called me "Bob!" Mentor that I am, I quickly straightened him out. Senior leaders often introduce themselves by name, not rank. It's a form of humility. But *never* take that as permission to use their first names!

What was this first impression? Negative, to say the least, and that's the way I saw him. Is that fair? Not necessarily, but it's true!

I hate it when I'm wrong, don't you? Luckily, I was only partially wrong. I asked him about the inspections, and he explained that there was a tremendous amount of work to be done to meet safety standards and that the shop was very disorganized. He had honorable and sincere intentions. He just hadn't read this book, so he didn't know about Habit 8 (you will soon). The poor guy was trying his best. He just needed some guidance.

I asked him about his language. "That's how my troops talk, Chief," he explained. You see, he may have been disheveled, but he wasn't dumb. (Notice how he quickly started calling me by my real first name, "Chief.") I gave him the sermon on how that type of language brings down the level of professionalism within a group. He understood and apologized. He was just trying to fit in. Actually, I think he was relieved to have my support for not using vulgarities.

After I gathered the NCOs, we talked about what I'd learned. They seemed to understand where the lieutenant was coming from. He'd just started on the wrong foot. Although he thought he'd impress them with his vision, making the unit better, the troops were unimpressed with their vision of him. He didn't look the part of a leader, so they didn't see him as such. Eventually, with some mentoring, that lieutenant became one of the wing's best. He took a different approach to leading that led to his troops respecting him and even following him. Was the troops' first impression still with them? Yes, but that lieutenant worked so hard to fix it that he never messed up as long as he was there. The troops never had to revert to their first impression.

Here's a War Story from a young NCO, TSgt Dalven Adams, very similar to the ones you're about to lead or are already leading….

As a young Airman, I was eager to learn the ins and outs of my job, but I was lacking that mentor to show me the ropes and guidance. I knew no one at my new duty station; Beale Air Force Base, but there was one TSgt that I gravitated to, and at the time I had no idea why. TSgt Derrick Longshore was an NCO that carried himself differently. One day, he asked me, "How can we set ourselves apart in the same uniform?" He left me with this challenging question, that I was not able to figure out. What I thought he meant was work ethic and drive, but those were not the answers he was looking for. He finally told me, "We all swear to the same oath but to speak for yourself, without saying a word; you must look sharp, clean, and be presentable at all times. In other words, he was telling me the first impression of what people see of me tells a story before even opening my book.

Fast forward a few years. I arrived at Elmendorf AFB to an Intelligence Squadron to perform my communications job, but little did I know that from day one of stepping in the place that I was being interviewed. The 373 ISRG was hiring internally for a Mission Program Assurance Officer, which was a Senior NCO billet but the commander at the time was not comfortable with the candidates. The position was responsible for overseeing the mission program and compliance for one group, three squadrons, one remote detachment, and a three-letter agency (NSA). So not only was this BIG but it was an ambassador for all those agencies, DoD, and the civilian sector. My arriving intent was not to work on mission systems or even be involved in the day-to-day ops. I was a comm guy coming to do comm things, now I was awarded the position as a SSgt with no intelligence background.

One day, after a meeting with my Commander and Chief, I asked, "Why me, what made me be the one out of all the NCOs and Senior NCOs in this building with Intel background?" They both looked at each other and said, "do you want to know?" In response, I said "Yes." My Group Commander said, "Well, from day one you showed up in this building with your uniform sharp from the blocked cap down to your cuff-stuffed bloused pants. It told a story about you without having to say a word. Looking at how you carry yourself told us that you respected yourself, but you also respected your service by wearing the uniform proudly, properly, and well beyond the standard. As a commander, we barely see NCOs taking care of their uniform (ABU) that way anymore, but to see it this day and age said you take pride in your service, and you are willing to take care of business beyond the standard." She trusted in me before knowing my name.

Chief said, "In my 22 years of service, I can say that I usually let people's records speak for them, but you let your appearance speak (first impression) and your records just supported you even more."

With their trust, I was able to bring the Group their first accomplishment of an Outstanding during the inspection and best in the NSA enterprise. To wrap this story up, your initial impression may be your only chance in many situations.

Starting Points

The first step toward developing a good habit starts with a good question. As Paul A. Samuelson said, "Good questions outrank easy answers." According to Naguib Mahfouz, "You can tell whether a man is clever by his answers. You can tell whether a man is wise by his questions." At the end of each habit, I'll provide you some questions—some starting points—that you should consider if you choose to develop the habit just discussed. I don't necessarily have the answers on specifically how you can improve, only because each situation calls for a different response. I will, however, provide you a way to reach me so that I can help you (see "Final Thoughts"). I hope these questions will point you in the right direction. Some wise guy once said something to the effect that the answers are always within the questions. Okay, maybe I just made that up. I have been called a wise guy before. Think! Question! Do! Be!

• **How would the person I want to be do what I am about to do?** (This is from author, Jim

Cathcart.) We're all about to do something. It makes sense to consider the effect(s) of our behavior BEFORE we act. That allows us to make better choices. It's always a choice. If you want to make a good first impression, think about what that looks like, choose to do it, figure out how to do it, then actually do it.

• **What am I doing that makes me look like a professional leader?** You may need to ask someone you trust to help you answer this one. We seldom know how we come off to others. Consider the opposite question, too. What are you doing that makes you look UNprofessional? Knowing the answers and making a commitment to professionalism is a good start to being a professional leader.

• **How am I striving to be humble in my daily interactions with my followers?** Remember that you're really the only person on earth who knows whether or not you're being humble. But remember, also, that others will always judge your humility by your behavior, not your intention. Not that you should try to impress others, but you WILL impress others. Be aware of how you express your humility.

• **What one thing can I do today, that if done in a professional way, will lead to my team's effectiveness?** This is a great question to ask yourself EVERY day, not just periodically. You're going to make an impression EVERY day. Maybe not a first impression, but your followers will be watching you and judging you every day. It may not seem fair, but it's the truth. Aldous Huxley said that "we shall know the truth and it shall make us mad." Don't get mad. Get better.

Words of Wisdom

Your talk talks and your walk talks, but your walk talks louder than your talk talks.
John C Maxwell

Your smile is your logo, your personality is your business card, how you leave others feeling after an experience with you becomes your trademark.
Jay Danzie

When someone shows you who they are, believe them the first time.
Maya Angelou

A thousand words will not leave so deep an impression as one deed.
Henrik Ibsen

Example is not the main thing in influencing others. It is the only thing.
Albert Schweitzer

Two things remain irretrievable: time and a first impression.
Cynthia Ozick.

It's not what you look at that matters, it's what you see.
Henry David Thoreau

To find out your real opinion of someone, judge the impression you have when you first see a letter from them.
Arthur Schopenhauer

HABIT 2

SHUT UP! LISTEN AND PAY ATTENTION

So, you're now a lieutenant! A leader of men and women! People look up to you. Okay, some people look up to you. As the leader, you and others think, for some unknown reason, that you possess knowledge and wisdom. So, they come to you for guidance. Since you are their leader and your purpose is to help your followers out as much as you can, you're more than willing to do so when a teammate comes to you with a problem.

"Lieutenant, I need your help," she says. "Aw, this is good," you think, "She respects me enough to seek my counsel." She just started talking and you're already thinking of something else—about how to fix her "problem." Your heart is in the right place, but your brain is already wandering, searching the recesses of your mind, trying to formulate an answer. You, like most of us, are good at jumping to solutions. The problem is that you don't know what the "problem" is. You aren't listening. You may be hearing—and that's questionable—but you're hearing with the intent to reply. That's not listening.

In Enlisted Professional Military Education, we teach Airmen a three-step communication process: tell them what you're going to tell them, tell them, and then tell them what you told them. That process works well if you're giving a briefing or teaching a class, but no one uses it to ask for help. Normally, the first part of the conversation is either background, icebreaker, or diversion. The punch line comes at the end of the conversation. If you stop listening as soon as you assume (and you know what that does) you know the question, chances are you'll provide a perfect answer to a totally wrong question. What's really bad is that since you know it's the perfect answer, you'll start patting yourself on the back early in the conversation and call it a success. In the meantime, your follower is still talking.

"Wait a minute!" you say. "I'll validate that my follower understands the answer I give her and that it does answer the question. I'll do that by asking her if she understands." Think a little bit. That follower is young and inexperienced. She's less experienced than you are. She came to you because she respects you. Surely, she had some trepidation—if not downright fear, mixed with awe—about coming to you in the first place. You ask her (with all sincerity, of course), "Do you understand?" What do you expect that youngster to answer? Maybe, "Nah, man, you're not even close to answering my question. Did you stop listening? Can't you do better than that?" I think not. She's going to answer, whether it's true or not, "Yes, Lieutenant!" As soon as she walks out the door, reality will sink in, and she'll wonder what planet you're from and why she came to see you to begin with. She'll immediately talk to her friends and pass on what happened. Word of mouth is the quickest and most effective advertising method. In about eight minutes, everyone in the entire unit will know that you're not a good listener.

Here's a question that will bumfuzzle all the men reading this. (Hey, I know because I am one.) What's the first step in the listening process? Nope. Try again. Men, it's *shut up*! Yeah, I know that hit you between the eyes. It should hit you between the ears. What's interesting is that the women get it on the first try! The truth is, we're scripted and trained to provide answers. Not to listen for the questions. Especially the higher you go up the ladder of success.

"Shut up" and "listen" are foreign terms to us men. No kidding! Say that to a guy and he'll say, "Huh?" But that's what you've got to do: shut up and listen for the question! Again, it's often at the end of the dialogue. Along with shutting up, shutting down is key. You have to shut down your own perspective to get the speaker's perspective first. Then you can use the benefit of all your wisdom to inspire the seeker to develop his or her own answer. Most people will already have the solution to their problem by the time they come to see you. They just don't like the answer or want someone to validate it. The best you can do is let them choose their own solution. Often, all they need is someone to listen. Do that, and the word will get out that you're an exceptional listener and leader.

Shutting up is the first critical part of the listening process; another part is paying close attention to the nonverbal communication. We often say more with our bodies than we do with our words. Tone of voice will tell you a lot about how people feel and what they want to communicate. Debbie, my lovely bride of 45 wonderful and fulfilling years, has exactly five ways she says my name. Two of them are good! Listen for feeling as well as content. That may take some skill, but you can develop it if you try. Take some listening courses if you can find them.

Listening, really listening, is a huge sign of respect. Not listening is a sign of DISrespect. To listen effectively you have to value and respect the person who's speaking. You don't have to agree with the person, but you have to respect and value that the person has their own thoughts, ideas, even feelings.

Listen with the intent to understand, not to reply. We, often, think that because we may have a little more experience than our followers we know better. "I've been there! Done that!" we think. NO, I HAVEN'T! I may have had a similar experience, but probably not. What I went through, even though it may have been similar, may have felt totally different. You and I are different, believe it or not, and how I perceive life is probably different, as well.

Practicing the L Word is critical to leading because it's the essence of communicating. You'll never lead if you don't, or won't, listen because real communication won't happen any other way.

Most schools that profess to teach leadership provide lessons on the critical leadership skill of communication, which is good. But how they perceive the communication process and what they teach is telling. How to express yourself. Yeah, you have to be able to give directions and share your vision and such if you're going to lead, but the real measure of how you lead will be in whether or not you listen to your followers. And not just for their words! The key isn't expression, it's impression. And the only way you'll be impressed by how someone feels is through listening for meaning, not just the words.

How many times have you asked someone how they're doing, and they reply that they're fine when you can see that they aren't? More than ninety percent of the communication process is nonverbal. It's about hearing with your eyes and your heart. It's intuitive.

"Don't judge me!" has become a popular phrase as of late. That's EXACTLY what empathic listening is about. NO JUDGING! Strictly understanding. Don't fix me! Just let me share how I feel and accept that I do, regardless of how you feel.

Empathic listening creates trust. More on that later. Trust, as you already know, is critical to your ability to lead. Shut up! Shut down! Just listen!

Strive to listen empathically. That means don't judge what you're hearing—just accept it as the speaker's view. After people tell you what they want to tell you, then ask if they're looking for advice. You'll have to use your intuition here. Remember that they may say what they think you might want them to say. Listen with your eyes as well as your ears. If they say no, don't offer it. You've done your job. If they do want advice, make sure they understand that it's based on your limited experience and that the final decision is theirs alone. Keep in mind that they might say yes when you ask them if they want your advice only because they think you want to advise them. As you become a better listener, you'll develop a stronger trust. When that happens, your followers will feel better about telling you whether they truly want your advice or whether they just want you to listen. When they come for the latter, count it a success.

As my fellow Chief, James Seballes, says, "Own that you don't know everything." You may be in charge, but that doesn't make you the smartest person in the unit. Chief Seballes would encourage you to ask questions for understanding, not judging. And, again, accept the answers, especially if they go against what you thought were the "correct" answers.

I'm sitting at my desk in the Family Support Center at Ramstein Air Base when Barb, our secretary, calls me on the intercom with some unusual seriousness in her tone. "Chief, please come down here ASAP," she implores. I do as she asks. As I walk into the welcome area, I notice a young lady sitting at a couch looking through a magazine I called the FIG. It was actually called a *Find It Guide*. The *Find It Guide* (FIG) included pretty much anything you ever needed to know about the Kaiserslautern Military Community. It was worth its weight in gold, and our Center distributed it. I could have sold those things for a lot more money than my salary.

Anyway, Barb immediately pulls me aside and tells me what's going on. The young lady had come into the Center ostensibly looking for a FIG.

After Barb had given her one, she chose to hang out. She wouldn't go away. Barb, who, by the way, was one of the most outstanding people I've ever worked with, had exemplary listening skills. She had an intuitive feeling that the FIG isn't why the young lady had come into the Center. And she was right.

After talking with her and enlisting the listening skills of one of our support experts, we came to realize that she was on her way to commit suicide. You see, people who get to that point are usually looking for help. If we're good enough listeners to understand that and listen with empathy, we may be able to help them find the help they need. After talking with her and empathizing with her, we took her next door to the Mental Health folks who were trained to help her. And they did. I saw her a few weeks afterward and she was doing okay. Thanks to Barb's listening skills she was okay. Again, listen for understanding.

One of the most important aspects of good listening is not interrupting. Have you ever talked with someone who finishes your every sentence? If they know what you're going to say, why waste your time saying it? Those folks can carry on your conversation and theirs at the same time. As difficult as it may be, don't interrupt. My lovely bride calls it interactive listening. It's become a cultural thing, I think. It's un-American not to interrupt! Let the speaker finish. Wait for that punch line.

Many American Indian cultures have a really neat listening process when they're in council. They have what they call a talking stick. The person who holds it gets to talk. Those who do not hold the stick get to shut up and listen. They say nothing at all until it's their turn to take the stick. My baby sister, Elva, gave me a talking stick for Christmas. I've had to adjust the process a little, though. I smack people with it every time they start to interrupt! (Hey, sometimes you do what you have to do! Okay, I don't do that, but I often consider it.) The point is, don't interrupt. It's rude, and you may miss the point if you do.

I'm at my desk, working on an award recommendation (a form 1206) when I have one of the most profound AHA Moments I've ever had! I've created the most powerful statement I've ever written, and I've written a lot. This sentence, alone, is going to ensure that my Airman wins the award that I'm recommending him for. THERE IS NO DOUBT! It's a done deal, I tell ya! Immediately after patting myself on the back, I commence to type it onto the form. Just five words in, Nancy enters my office without knocking, walks directly to my desk, and starts talking to me. In that instant, IT'S GONE! I hadn't finished capturing that incredible statement. I lost all track of what I had conjured up AND I didn't understand a word Nancy said because I was trying to remember what I was in the process of writing. Talk about a lose/lose situation! And you know how we retrace our steps when we forget something? Didn't work. I tried for hours and days. Nothing…. It WAS great, though, trust me.

Trying to be professional and a teacher, which I think all leaders must be, I, in as friendly a tone as I could muster, admonished her for entering my office without me inviting her in and talking to me, and, it was, by the way, TO me, before I was even prepared to listen to her. I did hear her, but all that did was distract me from what was way more important at that moment.

We expect people to listen to us, but don't always give them a chance to prepare. Listening takes preparation. The times I listen best are when I'm ready to listen. It takes preparation.

In raising two beautiful daughters, I always ensured that they knew I was available to them whenever they needed me. But you know when you tell pre-teens that, they hold you to it.

Elyse was eleven when she walked up to me while I was reading something very deep, maybe *War and Peace*, and started talking to me. It frustrated me so much that she had the audacity to think that I was going to stop what I was doing to listen to her. WAIT! I told her she could talk with me any time she needed to. I can't read and listen at the same time. The idea of multitasking is ridiculous. Humans aren't made for that, especially this one.

I put down my book and talked with her about how she might be more considerate of me when she saw that I was busy. "I want to listen to you," I told her, "but you have to give me a chance to prepare myself to listen. I'm not going to listen to you if I'm doing something else." She apologized and told me she understood and that she'd do better next time. I loved her so much. I can't tell you how proud I was of her for understanding. I got back to reading my book.

Fifteen minutes later, she returned. She didn't interrupt me. She lived up to her part of our bargain. But she just stood next to me, in eye's view, waiting for me to acknowledge her and invite her to speak. I'm not sure which was worse, the interrupting or the staring. But she lived up to our agreement! I did stop reading as soon as I got to a place where I could return without having to search for my place. I took a breath and listened to her attentively. It was a good conversation. We both won.

Prepare to listen. Don't just wing it. The first step in preparing to listen is to value and respect the person who has come to speak with you. Now, if that person hasn't made an appointment with you, it may be a bit difficult. You'll have to decide whether you can make the time immediately or make an appointment for later. If you make an appointment for later, be there at the agreed upon time and place. If you're not there, you're expressing disrespect. And, if you can, make sure you know what being respectful entails in the speaker's culture.

A colleague, who'd served as the deputy commander of the Air Force Academy Preparatory School a few years before our conversation, and I were discussing how people show respect to each other. We got to talking about how different cultures see respect, and respectful behavior, differently.

He went on to share a time when he and his commander had to dismiss a Navajo cadet because of her disrespect during an interview. What he said immediately caught my attention. A Navajo would be disrespectful? All of the Navajo I've ever known were very respectful. I asked him to explain.

He told me that the commander asked her to come to his office to discuss her bad grades and, seemingly, inability to accomplish the work required of students. "During the entire interview, she never looked up at the commander, never looked him in the eye, always kept looking at the floor," he told me. "The commander and I, obviously, took that behavior as apathy and disrespect. We dismissed her immediately."

"Sir," I replied. "Do you know that in the Navajo culture looking at people in the eye is disrespectful? It's a sign of aggression. The respectful thing is to NOT look people, especially an elder, which the commander was, in the eye. She did what her culture had taught her."

My friend's eyes teared up as he said, "Chief, we messed up, didn't we?"

"Yes, Sir, you did," I responded. "You dismissed someone's dream of becoming a professional Airman due to your ignorance of her culture and that culture's norms. She might have had the capacity to become the Chief of Staff of the entire Air Force given some guidance and understanding." He nodded and quietly left my office.

What happened happens more often than we know. As the saying goes, "We don't know what we don't know." It behooves us, as leaders, to know, or at least try to learn all we can about our followers. By definition and function, a leader is a leader ONLY because he or she has followers. People won't follow "leaders" who don't know about them, nor care to know about them, and their cultures. We live in a time where our teams include several, often many, cultures. An effective leader HAS to know something about each of those cultures. At least enough to ask the right questions about what's respectful and what isn't. The bottom line is that when you don't know, ask.

We would think that having the technology we have would make us better listeners, or, at least, help us to be so. I don't think so. I worked with a few folks who were always more interested in the technology, computer/iPad/smart watch, than me when I was talking to them. I can still see a couple of those folks looking at their computer monitor instead of me while I was talking. And each of them always rationalized that, "Oh, keep talking. I'm listening." BULL! I even said some preposterous things in those discussions that they didn't pick up on. I hope that hasn't happened to you. I really hope you haven't done that nor do it now. By the way, you know what you produce when you rationalize? Rational. Lies. Dr Stephen Covey taught me that.

TURN OFF YOUR TECH when you're about to listen to someone! And don't be looking at it, hoping that it will turn on by itself automatically. If the President wants to send you an email, you'll get it in due time. It probably won't require an immediate response. As Dr John A Kline says in his book, Listening Effectively, "Sometimes we don't listen because we are preoccupied. We have so many things to think about. Our mind is full of ideas, facts, worries. We are unable to put them aside while we listen. Nevertheless, good listening demands that we avoid preoccupation when someone is speaking to us." Turn your tech off and avoid distractions as best you can. Don't be looking around at what's happening around you. Pay attention to the speaker. In fact, if you can, find a place where there are no distractions.

Sit comfortably so that you can listen to the speaker and make your sitting position as "unconfrontational" as you can. Sitting across a desk from someone, especially if one is senior to the other is, to some degree, confrontational. There's a physical message that one of you outranks the other. That may be so, but if you want to listen for understanding you're not going to get the entire message if you're going toe-to-toe with a junior person. In the last office that I occupied I made sure I had a chair placed where I could turn away from my computer and almost next to where my guest would be sitting. I didn't want to argue, I wanted to understand.

Dr Kline admonishes us to ask, "How would I want others to listen to me?" Part of the answer to that question deals with what my friend and protégé, CMSgt (Retired) Lezlee Masson, suggests. Check your attitude. Your attitude will shine through your communication. Even online, on email. I remember having an email conversation with an NCO. After a few responses between us he asks me, "Chief, what did I do to piss you off?" I didn't know what he was referring to, so I asked him. He replied, "All of your responses are in CAPITALS!" I didn't realize that that's how you yell in an email. I do now….

Communication is key to success and even to having a good life. It requires listening, though, not just telling. Effective listening is done by listening with our hearts and eyes as well as with our ears. We have to be there. We have to see and feel what's happening to truly understand.

Shut up! Shut down! Listen! Pay attention! Be aware that once your followers know you're a good listener, they will seek you out. Very few people know how to listen. The power in that skill is that you'll be trusted, and you'll be able to help others. Listening will empower you to learn. And there is so much to learn, whether about people or the work you do. Listen and pay attention. Your followers will soon respect you for it. If you make it a habit to listen well, I guarantee you that you'll soon be on your way to becoming an Exceptionally Powerful Lieutenant!

War Stories

I'm sitting at the table on a Monday morning with all of the wing's group and unit commanders as the new general asks questions, trying to understand what his wing is about. "Why is that building across the street from the Security Forces building green when all the others on the base are beige?" he asks. Eyes in the headlights all around. Most of the commanders are new as well, so they don't know. No one responds with a plausible answer. The meeting goes on. I think nothing of it.

Later in the week, I make a point to explore the base a bit. Remembering the general's question, I go over to see that building he'd asked about. Sure enough, it's green - not beige like all the others around it. As I get closer, I notice a group of Enlisted Warriors in front of the building. Even from a distance, I can tell they're in an animated conversation. As I walk up, I quickly get the gist of their discussion and ask the senior person what's going on. "We have to paint this building by the end of the day, Chief," the NCO in charge tells me. "Why?" I ask, "It seems like the paint is still holding up." "Our commander said the general doesn't like the color, so we're going to paint it the same color as the others," he responds.

Snitch that I am, I immediately go back to headquarters and talk with the general. "Sir, did you tell anyone to paint that green building across from the Security Forces building?" "No,

Chief, I didn't. Why do you ask?" "Because there's a group of Airmen over there getting ready to do just that. And they've got plenty on their plate right now without having to be painters."

Needless to say, the general gets on the horn (that's a tele-phone term for you young folks) and stops that tasking. "So what?" you ask. Here's the listening lesson: that group's commander had read into the general's question much more than was there. As the general and I talked, he made it clear that his question was a sincere and naïve one. He just wondered why that building was a different color. The commander assumed (there's that word again) that the general didn't like it and being an overachiever (as many commanders are), he was trying to please his boss. He didn't shut up or shut down; instead, he started coming up with a solution for a problem that wasn't even a problem.

Do you think that the commander's over-aggressiveness might have had an effect on the followers? A big one! They had other, much more important, duties to accomplish. That commander was more intent on doing what he thought his boss wanted than considering what was right in the larger scheme of things and doing that.

Listen. Pay attention. Make sure you understand the message. Don't assume—it makes an ass out of you and me. You probably don't need any help doing that!

Here's a WAR Story from a prior-enlisted officer, Major Lisa McCarthy….

Though not my best moment as a human or leader, I wanted to share an example of what not to do, where I failed miserably, lacked total empathy in many ways, and was too focused on the person, their failure to uphold military standards, and blatantly missed the real issue.

As a second-year lieutenant, I was serving as the Materiel Management Assistant Flight Commander and had been in the seat just a mere few months by this point. I had come into the Flight with a preconceived understanding of specific individuals, based on my predecessor's feedback, and had used that as a foundation leading into this particular situation. We had a female NCO who had a track record of failed PT tests and other performance issues. She was a relatively quiet individual and had been categorized as a poor performer overall. During this period, she had reported a sexual-assault-related incident where she was the victim. Her incident had gone up through the SARC and command channels and was being handled well above my level, where Flight leadership served in more of a support role.

However, her performance had continued to dwindle and further resulted in a PT failure. I remember her coming into the office upset. I don't quite remember what exactly was said between us, but I remember we had an audience (Flight Chief and Flight Superintendent). She had come to share information with us, and I remember immediately dismissing her, making a spiteful comment about her PT failure, and leaving her with the impression that I did not believe her situation, nor did I care about anything past her PT failure. She left with clear disdain that I lacked total empathy for the situation she was going through.

Later that week, the First Sergeant came to the Flight and shared where I was in the wrong and that the individual was on the verge of going to the IG but decided not to pursue it. He, the rest of the Flight leads, and I spent lengthy time discussing the situation and how I needed to re-vector myself and be the support system our Airmen need, especially when they are going through life situations where we may be the only support they have.

I walked away learning a lot about myself that day and what type of leader I needed and wanted to be for my Airmen. Moral of the story...I was at total fault in this instance and failed to be empathetic, hear her out about why she was upset, and, ultimately, failed to recognize the repercussions of her situation that were the root cause behind her poor performance.

I later went back and apologized to that individual, but by that point, I had already lost her trust and confidence in me. Soon after, she left our unit and pursued her planned separation. Since that situation, I made a promise to myself and to God that I will care for my Airmen in the way they deserve. I will show empathy as warranted and be intentional in making them feel heard and seen. Though I am still imperfect, I make more deliberate effort toward continuing to grow in this area.

Starting Points

• **Do I really want to understand, or do I just want to impress?** As you already know, empathic listening is about understanding, not judging. It's sometimes difficult not to judge, especially when the issue elicits emotions. You have to control your emotions in order to listen empathically. It takes practice. Practice often.

• **Do I shut down to hear the speaker's message from his or her perspective, not mine?** Simon Sinek says that people want only two things; to be valued and valuable. Empathic listening begins with valuing the speaker as an individual with their own story. You haven't been through what the speaker has been through. Value that his/her story if valid and theirs.

• **Am I observant of body language?** Some really smart people say that up to 97% of the communication process is no-verbal, that means a lot of the message isn't delivered by words. Pay attention to how the speaker says what he/she says. Use your intuition to guide you toward what the speaker is really trying to convey.

• **When should I give advice?** ONLY when it's solicited. Offer to give it but remember that it may not be needed. That's okay. And if your offer it and/or it's requested, it'll be from your experience, which may or may not be valid.

Words of Wisdom

Courage is what it takes to stand up and speak; courage is also what it takes to sit down and listen.
Winston Churchill

I have two ears so I can hear both sides of any argument.
Noah benShea

Ideal conversation must be an exchange of thought, and not, as many of those who worry most about their shortcomings believe, an eloquent exhibition of wit or oratory.
Emily Post

The word 'listen' contains the same letters as the word 'silent'.
Alfred Brendel

The most important thing in communication is hearing what isn't said.
Peter Drucker

Listening is about being present, not just about being quiet.
Krista Tippett

We have two ears and one mouth and we should use them proportionally.
Susan Cain

One of the most sincere forms of respect is actually listening to what another has to say.
Bryant H. McGill

BOB VÁSQUEZ

HABIT 3

LOOK UP!
ATTITUDE IS EVERYTHING

"He's got an attitude!" Have you ever said that about someone? Has anyone ever said that about *you*? (Okay, if you're a she, has anyone ever said, "She's got an attitude!" about you?) Chances are you're nodding your head or looking to see if people are around so that they don't see you acknowledging the truth. *You've* got an attitude! Yes, you do. As I do. And mine is probably bigger than yours. After all, I've got more to have an attitude about. The question is, what *kind* of attitude? In case I haven't said it yet, life is all about choices. You will choose—and express—either a positive attitude or a negative attitude. There is no in-between. You have to choose, so make the *right* choice. What's the right choice? Imagine the following:

"That person" comes into the unit every day with a scowl on her face. Of course, we're talking about someone else. It's more convenient that way. On this particular day, the planets and stars align, showing you the answer to the most important issue you're working on at the unit. It's the one change that can make everything—and I mean everything—better for everyone. HEIRPOWER! When implemented, this idea will produce quantum progress. You happen to work for "that person," so you can hardly wait to share this great thought with her. However, comma, after making your pitch and true to her normal attitude, she shoots down everything—and I mean everything— you say. She doesn't even hear you out (she hasn't read this book, hence doesn't know Habit 2) before she starts ranting about how stupid an idea it is. I know that's never happened to you, but pretend it has.

Now imagine going through the same scenario with a boss who HAS read this book and is applying its tenets. After hearing your plan completely, without interrupting, she looks you in the eye and says, "Great idea! Let's see if we can make it work! It may not work exactly as you've envisioned it, but we can tweak it along the way until it works. We'll all be better for it. Great work, Lieutenant! Thanks!"

Okay, now choose. With whom would you prefer to work? Do you need any hints? Probably not. Do you think that the attitude a leader expresses will affect those whom he or she leads and works with? His or her attitude will permeate everything. ***Attitude is everything because it affects everything***—and I mean ***everything***! "Yeah, but…," you say. "Things happen. Especially on the way to the job. I can't help that, and that affects my attitude!" you're rationalizing (refer back to what I said before about rationalizing.) Well, you're right… kinda…. We are affected by external events. And those events CAN affect our attitude, if we choose to let them. I think I already told you that everything is a choice.

Let's see how smart you really are. Where do people's attitudes come from: (a) the way they were raised, (b) the responsibilities they have on their shoulders, (c) their DNA, or (d) the choices they make? You *are* smart! And growing smarter as you read, right? The truth is that **attitude comes from the heart, works its way through the brain into the mouth, and comes out through the hands.** Kind of like the leg bone is connected to the knee bone. . . (Ask some old*er* person to share that song with you.). Your heart is where you keep your purpose. Why are you doing what you're doing? If it's for money, forget it! You'll never make it rich in the lieutenant business. If it's to support your followers and help them grow, you're on your way to becoming an Exceptionally Powerful Lieutenant! Your most sincere intentions, however, can become misguided if you don't think about how to make them real to others—namely, those you lead. You have to think about their needs as well as how they communicate and then try to speak at their level, which may be higher than yours. Remember the old adage "Sticks and stones may break my bones, but words will never hurt me"? Isn't that *stupid*? Actually, sticks and stones may break our bones, but words will break our hearts and spirits. What we *do*, with the wrong attitude, can cause even more irreversible damage. ***Work on saying and doing the right thing at the right time in the right way for the right reason.*** You'd expect that from others, wouldn't you? *Your attitude is yours and yours alone*. Choose the right one. You *will* share it—like it or not.

Here's an essay I wrote for my *PowerPact Leadership Lessons!* blog, titled *Garbage In, Garbage Stays!*

Back in the day…way back in the day, we had a saying that is still true, maybe more true, today. That saying was, "Garbage in, garbage out." In other words, whatever we choose to put into our brains produces similar results through our thoughts and our actions. If you put negative thoughts in your brain, you'll produce negative results. It's a natural law.

I'm bad at reading all the posts on Facebook. I'm trying to quit. I've embraced social media, in particular, Facebook, because it's how I stay in touch with great people I've met along the way, and some I haven't, but seem to be good people. What interrupts the positive flow is all the garbage that I can't seem to delete or keep from returning, like ads and news stories. I'm blessed with Friends who post positive thoughts and stories that offset or counterbalance the negativity, the garbage.

I served in the United States Air Force for 50 years. Although it was an evolutionary process, I was committed to protecting every American's rights even if I didn't agree with them. What bothers me most about all the negativity I see around me is that those who post such garbage never provide solutions. Yes, things are bad in some respects, but not ALL respects, but how do we make them better? More importantly, how do I make them better?

What I propose is that garbage in, garbage stays. When we don't produce solutions in our minds that transfer into better behaviors by our hands, we'll keep producing garbage, which, it seems to me, is a lot easier to do, but totally ineffective.

You choose your attitude. It's influenced, at least in part, by what you put into your brain. And the more garbage you accept, the more it'll remain in you until you throw it out, which will, again, affect your behavior.

So, my personal commitment is what it's been for decades. My Brother Chief, Kenny Mott, recently posted a thought on Facebook that embraces my commitment, "Today, I will make someone's life better." I'm going to ensure that everything I let get into my brain moves me toward that end. No more garbage! I'm going to make a positive difference today! Will you?!

Let me share an analogy I borrowed from Lt Col (Retired) Dave Keller, a good friend and mentor. What is this? Oh, right, you can't see that I'm holding a thermometer. Pretend you can see it. Again, what is this? Yes, that's right; it's a thermometer. What does a thermometer do? Okay. Yes. It assesses the environment, doesn't it? You were pretty close. A thermometer tells you whether it's hot or cold or how hot or cold it is, right? It assesses the environment. Now what is this? Oh, sorry. Pretend you see the thermostat I'm holding. Yes, that's right. It's a thermostat. Now, what does *it* do? It *controls* the environment. You're right! Okay, don't get too technical; just go with me on this. (Some of you are *too* smart!) Now here's a choice question: would you prefer to be a thermometer, an instrument that assesses the environment, one that goes around telling everyone when life stinks; or would you rather be a thermostat, controlling those external things that bombard you and can influence you? Remember the two scenarios at the beginning of this lesson? Which of those two bosses was a thermometer? Which was a thermostat? Which are *you*? The question I asked you earlier was "with whom would you like to work?" How do you think your followers would answer that question? Probably the same as you would.

Yeah, I know. Some of you are saying that's a faulty analogy. Blame Keller; he's the one who showed it to me! (Okay, I'm just kidding; my attitude started showing! Don't blame anyone.) Here's the deal: although a thermometer doesn't *always* say it stinks, it always takes its attitude from external sources. The thermostat controls those sources. You may think like Geraldine Jones, Flip Wilson's character, that "the devil made me do it!" (Okay, find that old*er* person who helped you with the knee-bone song and ask her to tell you who Flip Wilson is.) **Your attitude comes from you.** You may let external events influence it, but you will always have to make the choice. You choose your attitude and then share it with everyone around you. The sharing part is not a choice. You *will* share it with everyone around you because an attitude is contagious. It affects everyone you touch. It even affects morale.

Let me talk about morale for a minute. We often think of morale in terms of your followers being happy. Your followers are *never* happy! Okay, that's not true. Sometimes they are. They're happy when they have good leaders to follow. Being happy is not necessarily what morale is about though. It's about attitude! I've been in places like Bosnia, Turkey, and Saudi Arabia, where the followers weren't happy, but morale was high. I remember being in the mud, literally, in Bosnia, working alongside soldiers who put their lives on the line every day. They certainly weren't happy, but morale was high due to their attitudes, which were positive because they realized that what they were doing had meaning. They had to do their jobs because their buddies' lives depended on it. They had a sense of purpose and mission. They refused to let external things keep them from performing their duties. They were thermostats! Every time I think of those men and women, my heart swells with pride. They were—and still are, I'm sure—warriors of honor! Big props to each of them! ***Morale has to do with attitude. A positive attitude, especially from leaders, will filter through a unit and create high morale.*** It always has. It always will.

Earlier, I referred to an idea that would produce quantum progress. Chances are that you understood that to mean huge progress, which is how most of us perceive the term *quantum*. The dictionary, however, defines it in terms of smallest, not largest. I used that term intentionally. Oftentimes the actions, the behaviors that have the greatest impact are the smallest and easiest to generate. That small thing you do creates big results, and those big results create small, high-impact behaviors. And then the cycle starts all over again, in a circle. Think about this: what very small, inexpensive act could you do daily that would produce a huge, positive impact on your followers? Something you can do every day with almost no effort but will make you a much stronger leader? An action or behavior that will create a positive attitude because it *comes* from a positive attitude? I won't lead you on too long. This isn't rocket surgery. Simply saying "good morning" to your followers—and meaning it, of course—will reap *huge* benefits! I'll explain why in a second. What would that cost you? How much effort would it really take? Very little. What effect would it have on your followers? A quantum effect. Quantum in the sense that such a small investment can affect everyone in the unit and beyond. (I'm beginning to sound like Buzz Lightyear in *Toy Story 2*, aren't I? You may need a young*er* person to help you with that movie.)

It's the little things you do that express your attitude. Here's why those small acts produce such large benefits. Everyone wants to feel valued. Even lieutenants. When you acknowledge a person's presence (or is it *presents*?), you're telling that person you care. (The next habit focuses on caring, so I won't elaborate here.) Every person in our military forces is valuable. Exceptionally Powerful Lieutenants will ensure that all of their followers know that they're valued every day. Wishing your followers a good morning and acknowledging their presence is a very small price to pay for the return you'll get. Add a smile, and they will say great things about you. In fact, a sincere smile and respectful greeting will say much more about your attitude than all the speeches you can ever deliver. (For those of you who work different shifts, you'll have to say the appropriate words, but I think you catch my drift, don't you?)

When I was young (I was, many years ago), my mom and dad would always wish my siblings and me a good morning and ask how we slept. That may be a cultural thing; I don't know. First, they wanted to welcome us to another wonderful day. Second, they were sincerely interested that we'd had a good night's sleep. That was a great start to every day! We didn't have much growing up in materialistic terms, but we had all we needed in attitude. That's where I developed my "Isn't this a great day?" war cry.

I often hear people tell me, "Well, I'm not a morning person." Get over it! I think that's a cop-out for choosing a negative attitude. Imagine the power of developing a habit that puts you in a positive frame of mind at the start of the day. Do you think it might carry on the rest of the day? Do you think it might affect the people you lead? Try it, and then answer those questions. Try what? Think about the next couple of paragraphs.

Although I chose "Attitude Is Everything" as the subtitle for this habit, the title is "Look Up!" I'm a self-acknowledged hypocrite, so I'd like to practice that now. Although attitude is everything, gratitude is more. (Okay, I'm going in circles. I like circles, especially when I'm in the right ones.) Here's how I see it (and it *is* a circle). Your attitude will come from a sense of gratitude that will affect your attitude. If you're truly grateful for what you have, your attitude will reflect that. And it will affect everyone around you.

The mother of Zig Ziglar, a motivational speaker, said, "When the outlook isn't good, try the up-look." In other words, look up. Search for meaning beyond you and above you. Your attitude will reflect your hope. Hope is a very powerful concept, particularly when you're leading. Your business is war, whether you like it or not. As a leader, your followers will often look up to you to gain their hope. When you provide it to them, they will excel because their attitude will be positive. Hope is like faith. You can't hold it in your hands, but you know when you have it and when you don't. Your followers do, too! It's interesting that we find hope in ourselves by having faith in something larger than we can possibly be. Find faith and hope and let them raise your attitude so that every one of your followers will say, "I'll go to war with that Exceptionally Powerful Lieutenant! Any day! Any place!"

Chaplain (Retired) Chappy Watties, my pastor and good friend, and I were sitting in Bosnia waiting for a plane to take us back to Germany. We're both philosophers although he's much wiser than I. As we were sitting there analyzing the world around us, he told me, "Chief, there are two types of people in this world. Those who get up in the morning and say, 'Good morning, God!' and those who get up in the morning and say, 'Good God, morning!'" Again, it's your choice. Attitude is contagious, and your followers will catch it from you. You choose.

Be a thermostat! Choose a positive attitude! Be thankful for what you have. Share that gratitude with your followers. Stand beside them and help them look up and aim high.

War Stories

I'm at the Brandenburg Gate one cold and dreary October morning in Berlin, Germany. (Okay, almost every morning is cold and dreary in Berlin. So, what's new?) I'm sitting in a bus full of enlisted Air Force Bandsmen waiting to march in the first-ever Unification Day Parade. I'm serving as the drum major for the United States Air Forces in Europe (USAFE) Band. The troops are restless. They're cold, and it's raining. You can probably understand how they feel only if you've been there. As troops sometimes do, they start to get rowdy. Some are "tough" enough to approach me and ask whether we can cancel our march because "the cold, wet weather can negatively affect our instruments and our uniforms, which would cost a lot of money to replace." Any warrior would know the answer. Let me say that these are real warriors. It's just that after being out all night in Berlin (acting as American ambassadors of goodwill, of course), it's very difficult to think clearly first thing in the morning.

Anyway, being the astute student of group dynamics that I am, I stand up and start trying to inspire the troops by telling them how important this parade is, politically and emotionally, not only to the German population but also to the United States government. I was *on*, man! I could see morale growing with my every word. I finished my short-yet-very-eloquent speech by saying, "It could be worse. It could be snowing!" As I said that, their eyes changed as they looked beyond me. Someone pointed. I turned around to see that it *was* snowing. We still marched in the parade.

Regardless of how tough things get, they could always be worse. If they're not worse, be grateful for it. By the way, if you'd like to see a picture of that parade, go to the USAFE Band's Web site. You have to look hard to see me since I'm facing the band, but you can see the cold, wet troops marching proudly in front of the Brandenburg Gate. They represented you well... after they adjusted their attitudes.

Here's another perspective from one of my favorite protégés, Senior Master Sergeant Mealinda Koory.

There I was, a young Staff Sergeant, TDY on a great opportunity for networking and learning. As a Space Operator, it wasn't the norm to be at an Intel conference, but I had been selected to represent my squadron as an assistant at the conference.

Day one, smooth sailing, prepping and setting up for the conference attendees. I even got some free time to run the trails and swim some laps in the warm sun of the South.

Day two, uh-oh. The Commander requested we be at the squadron at 0600, ready to work. And I was! But she wasn't. In fact, she didn't make it until almost 0900. For nearly three hours we sat around—with me complaining. Here I sat in the midst of some pretty smart people, offered an incredible opportunity to learn and network, but I couldn't see past the inconvenience of having to wait around. Looking back, I can only imagine what they were thinking. I often wonder why someone didn't tell me to shut up and look on the bright side. That would have been what my husband and I call our billboards.

Sometimes we get stuck with blinders on, unable to see the positive side of our situation and we need that billboard to smack us in the face. This is the attitude check. Have you had one recently? Just yesterday I had one while shopping for Thanksgiving. There I was, checking off every item on my list but annoyed by the *thousands* of other shoppers out with me (no, I'm not exaggerating, have you been to Walmart the week of a holiday???). Like I said, I was checking off my list with little hassle, but I was annoyed that I was at the store the week of the holiday. What I didn't focus on were the facts that a) I was filling a cart with tons of food that would feed and nourish my family, b) I had plenty of money in my account because c) I'm blessed to hold a fantastic job that provides for my family and me. It took me a minute to readjust my mindset and focus on the blessings instead of the small annoyance of a crowded store. My ATTITUDE determined my outlook, instead of the other way around. Thankfully, some really smart people have helped me instill some internal billboards that turn on nearly all the times they need to, and my attitude gets adjusted by focusing on the good in my life. I encourage you to do the same and look at every situation as an opportunity, no matter the 'inconveniences' they may bring with them.

For those who may be reading and knew me at an Intel conference long ago, I apologize for my young, misguided ways! I appreciate the opportunity I was offered to learn from some of the smartest watchers out there.

Here are a couple more musings from my *PowerPact Leadership Lessons!* blog.

AN ATTITUDE OF GRATTITUDE

The best attitude an Effective Leader expresses is an attitude of gratitude. It will permeate all of your actions and your relationships on a daily basis.

I create my attitude of gratitude daily. As I awaken every morning, the first thing I do is give thanks to the Creator for all of my blessings. I give thanks for...

This country I'm blessed to live in. Yeah, it's not perfect, but I've lived in other countries and visited many, as well. None of them compares to the United States of America. NONE! As John C Maxwell says, "There's no place like this place anywhere near this place, so this must be the place." We've still got plenty of work to do to make it what we, as a nation, want it to be like, but we're better off than any other country in the world. There are many people willing to die to come here. No other country can boast that. I'm grateful for my country!

The people I'm blessed to live and work with. I'm honored to have been able to serve with the best people in the world on a daily basis for the past fifty years. Chiefs don't cry, but sometimes our eyeballs sweat. Mine sweat when I hear and read about the great accomplishments My People achieve. They also sweat when we lose one of our own. I can't help but swell with pride when I think of my nuclear and extended families as everyone grows into who the Creator sent them here to be. I'm grateful for my family!

The opportunity to produce HEIRPOWER!, which includes you! I'm so blessed that you're willing to make time to read my thoughts. My purpose of producing HEIRPOWER! grows exponentially as you share your thoughts and yourself with others. I'm grateful for you!

My health and my faith. One day you'll understand. As you evolve, you'll begin to hear noises and feel aches that your body emits without your consent. At first, you'll wonder where they're coming from and, eventually, you'll accept them as yours. Despite that, you'll learn to be grateful for having awakened. The key is not losing your awareness. Especially of the blessings the Creator bestows upon you. I'm grateful for me!

These are what I'm grateful for on a daily basis. I don't deserve all I'm blessed with, but I count it all as Grace. I commend you to consider your blessings and include them in your attitude development. One of my most favorite authors, Jim Cathcart, asks, "How would the person I want to be do what I'm about to do?" The answer to that question begins with your attitude. Today, before you go out to take on the world, take a few minutes to develop the right attitude…an Attitude of Gratitude!

HUGE IMPACT

Almost every conversation I had with others this week, focused on the question, "How do leaders create a positive impact?"

Having served in the military for 50 years, I'm accustomed to acronyms, so here's one I created that's based on my perspective of how leaders create a positive impact. It's HUGE!

Be **HUMBLE!** That can be difficult because humility is difficult, if not impossible, to measure. How do you know if someone is humble? What's the criteria? The official definition of humility is "a modest or low view of one's own importance." It seems to me that you're the only one who knows whether or not you're being humble. You're really the only one who can make that assessment, but will your followers know? That's who you're trying to impact, right? Maybe they'll know. Maybe not. See how difficult it can be? But here's what you, as a leader, can do to express your humility. Practice being modest, respectful, considerate, and unselfish. This is the way. (Yeah. I'm a fan.)

Be **UPLIFTING!** As Robert Ingersoll said, "We rise by lifting others." Zig Ziglar said, "He climbs highest who helps another up." And, finally, Maya Angelou said, "Try to be a rainbow in some's cloud." Whoever you're intending to lead today can use a pat on the back, a compliment on their work, even a word or two of encouragement that will empower them to be the best version of themselves. They're great! Tell them so. It'll cost you nothing. But the impact you'll have will be powerful.

Be **GRATEFUL!** Create an attitude of gratitude. Be grateful that you get to lead each of your followers, as challenging as that may be some days. But that's what's cool about being a leader, overcoming the challenges. Being grateful will have an impact that you may not readily see, and it will be effective beyond your knowledge. I make sure I count my blessings, which are many, when I first wake up in the morning. I live in the greatest country in the world. I work with the best people on Earth. I woke up healthy! I woke up! I have a lot to be grateful for. I bet you do, too. Be grateful. It's contagious.

Be **EMPATHIC!** Empathy is one characteristic we're in dire need of today. Empathy is about understanding, not judging. As Stephen Covey taught me, "We can agree to disagree agreeably." We actually can! I don't have to agree with you to still care about you and respect you. That's how you feel about the subject. I don't agree, but I respect that you feel that way. We're good. And that's good. Dr Covey also taught us to "Seek first to understand, then to be understood." The more we practice that, the more impact we'll have on ourselves and on others. Leading effectively starts with me practicing understanding and not judging.

These four, seemingly simple, ways of thinking and being will have a HUGE impact on you, your followers, and our world. Think about how you can practice them today. Just one behavior in each category, and I guarantee you HUGE results!

Starting Points

• **Isn't this a *great* day to be an American Warrior?** I've been asking that question of audiences for decades. I ask that of myself ever morning. It guides me toward being grateful to be able to serve my great country next to great people. I commend you to ask yourself this question. The question itself may generate an attitude of gratitude within you. Then, pass it on....

• **Have I acknowledged my followers' "presents" today?** My Chosen Brother and mentor, Chief Master Sergeant of the Air Force (Retired) Dave Campanelle taught me something that paid off highly. When I became the Senior Enlisted Advisor to the Commander of the 92nd Air Refueling Wing, he told me to take a map of the base, note where my followers worked, and systematically visit each building. Once I'd visited every building start again. I made all the rounds a few times while I was at Fairchild AFB. I pretty much wore out that map, but it kept me on track. You can do a similar thing to ensure you acknowledge all of your followers. It may be a smaller scope, like the offices in your building, but it will have the same effect.

• **What am I truly grateful for?** You're probably blessed more that you think and maybe even take for granted. I know I am. But I don't. Take my blessings for granted, that is. As I said before, I wake up every morning thanking my Creator for all of the people I live and work with and for all that I have. I don't deserve all that I'm blessed with, but I count it as Grace. Thinking about the good in our lives makes the not-so-good better. Try it. You may like it.

• **Am I being a thermostat today?** You can't control everything in your life. But you CAN control your attitude. And by doing that you'll affect the attitude of others. I'm going to assume that you want to make your world, and your followers' worlds, better. Do it! You have that power. As Gandhi once said, "Be the change you want to see in the world."

Words of Wisdom

A little spark kindles a great fire.
Spanish proverb

*The soul that gives thanks can find comfort
in everything; the soul that complains can
find comfort in nothing.*
Hannah Whitall Smith

*Growth in wisdom may be exactly measured
by decrease in bitterness.*
Friedrich Nietzsche

*All I have seen teaches me to trust the
Creator for all I have not seen.*
Ralph Waldo Emerson

*That man is happiest who lives from day to
day and asks no more, garnering the simple
goodness of life.*
Euripides

When I count my blessings, I count you twice.
Irish blessing

When you arise in the morning give thanks for the food and for the joy of living. If you see no reason for giving thanks, the fault lies only in yourself.
Tecumseh

This is a wonderful day. I have never seen this one before.
Maya Angelou

Habit 4

Be Care-Full!
Take Care of Your Followers

LEADERS CARE
(from my *PowerPact Leadership Lessons*! blog)

Decades ago, I studied with a professor who taught me that "Life is simple...once you understand its complexities." Profound thought, if you think about it. Similarly, leading is simple, once you understand the complexities. Leaders CARE! That's it! If you do, you will! CARE and lead, that is. Here's what leaders do.

Leaders are COMPASSIONATE. Two key complexities of being compassionate are being there and being empathic. When you're NOT there, whether physically, mentally, emotionally, or spiritually, you're not leading. You can't! When you ARE there in those four areas, you'll have the opportunity to lead. You still may not accept the opportunity, it's always your choice, but at least you'll have that choice. Leaders always accept the opportunity. Being empathic requires us to be understanding. That's all! NO JUDGMENT! That's the hard part. You have to accept your followers for who they are, not compare them with who YOU are.

Leaders ACKNOWLEDGE presence and presents. You can't imagine the power of greeting someone, especially eye-to-eye. The eyes don't lie. They express compassion, or not. Letting someone know that you know that they're there is critical to being an Effective Leader. And expressing to them that they're valued because of the gifts they bring to the team will last way longer than the expression itself.

Leaders RESPECT their followers. On THEIR FOLLOWERS' terms. We often attempt to show respect on OUR terms. That doesn't always work. How do you show respect? Ask! "I want to refer to you as (fill-in-the-blank). Is that okay? I want to be respectful. Please tell me if it's not." Once you have permission or guidance on how that person sees you as respectful, practice it until it becomes habitual.

Leaders EMPOWER their followers. Okay, no one empowers anyone. That's a myth. People empower themselves. Telling or making someone do what you want isn't leading. It's coercion. Inspiring and equipping your followers to empower THEMSELVES to be their best selves is empowering. When you do that, people will follow you.

See how simple this all is? The complexities are practicing these key components in a consistent way, and with the INDIVIDUALS you intend to lead. We often think that a title or rank or status makes us leaders. NOPE! It doesn't work that way. The ONLY thing that makes a person a leader is if he or she has followers. Nothing else! If you do what I just suggested, you'll have followers, which will make you a leader. It's that simple. Leaders CARE!

Man, I loved that guy! Well, actually, I still do. I worked with a general officer with whom I would still go to war without hesitation. If he called me this very minute and said, "Chief, I need you to go to Timbuktu with me," my only question would be, "When do you need me to be there, General?" As much as I truly found it an honor to serve with him, he also frustrated the dickens out of me. He was never—and I mean never—on time. His wife is laughing her head off as she reads this because she knows how bad he was about being late. She often said, "He'll be late for his own funeral."

Let me tell you that he never intended to be late. Oftentimes, he got so deep in discussion with people, trying to help them out, that he'd forget he had another appointment. Remember that I told you this: when you become a general officer, you will continuously run from one appointment to the next. Unintentionally, the message my general sent to those whom he kept waiting was that he didn't respect their time. As hard as I tried to explain to audiences that he meant no disrespect, that's how they often took it. Know this: what you intend doesn't matter nearly as much as how others perceive your behavior. Perception is truth to the perceiver. If those you lead think you don't respect them, you'd better change your behavior because it's the only thing that will change their minds.

Leadership guru John C Maxwell says that followers have to buy into their leaders before they buy into their vision. I believe that's true. How do leaders get people to buy into them? They have to show that they care. How can *you* show that you care? More on that in a minute. You may have heard the adage "People don't care how much you know until they know how much you care." The hard truth is that people *do* know how much you care… by your attitude and what you're willing to do to help them empower themselves. If you care for them and treat them like winners, they'll live up to your expectations. On the other hand, if you treat them like dirtbags, they'll live up to your expectations.

Caring is about the little things you do, or don't do, that express your concern for your followers—or your family, for that matter. I mentioned the little things in the last habit. To be care-full means you have to be full of care. The subtitle of this habit, "Take Care of Your Followers," is critical to your success. Exceptionally Powerful Lieutenants will *show* their followers that they care. You'll expect your followers to be on time, won't you? It's a matter of respect, isn't it? Here's the most powerful lesson in this book: **to gain respect, you must give it.** That's the *only* way you'll gain it. Trust me! What you give, you will get in return. That's the way of the world. It's one of life's paradoxes. If you respect your followers enough to be on time, they will do the same. Interestingly, one of the biggest issues we dealt with in that wing led by my general was on-time takeoffs. Could there have been a correlation? You decide.

I was honored to work alongside an Exceptionally Powerful Colonel during my time as the Senior Enlisted Advisor to the Commander of the 92nd Air Refueling Wing at Fairchild Air Force Base right outside of Spokane, Washington. One of my favorite people and officers I worked with was the Wing Vice Commander, Colonel Wally Dill. He knew everything, and I mean everything, about running that wing. He'd forgotten more about it than the rest of us would ever know in our lifetimes.

Colonel Dill would often invite me to go out to the flightline with him to time our team's their take-off times. I recall one morning as we sat in the dark in his official vehicle at the end of the runway with a clipboard in one hand on which the morning's list of chocks was attached (We didn't have electronic devices for that in those days. We didn't even have smart phones!) and a stopwatch in the other, waiting for the morning's flights to start taking off. As each flight took off, I was to note the time it'd taken for "wheels up" and mark that information on the sheet. "How did that one do, Chief?" the good colonel would ask me. "Right on time, Sir," I would respond…most of the time. "How was that one, Chief?" he knew it wasn't on time. "Ugh, about five seconds late, Sir," was my reply. "Mark them up, Chief," he instructed me. "Sir, with all due respect, they were just five seconds off. You really want to write them up for that?" I pled. You see, after we were done out there, we'd report to the entire staff, at our weekly staff meetings, the results of each day's on-time departures. No one wanted their name to appear on the board as being a late take-off.

"Chief," Colonel Dill looked me in the eye. "The planes our folks are going out to refuel expect us to be there on time. Actually, we have to be there early. They may be almost out of fuel by the time we rendezvous with them. We have to be there when they get there or they could, literally, fall out of the sky. Now, do you want to be responsible for a plane falling out of the sky because we were five seconds late to their refueling time? None of us wants that, so we have to make sure we're on time EVERY time. Five seconds may not seem like a lot, but it could be critical, and being on time is a cultural norm. We're NEVER late. We're ALWAYS early. That's what we want to be known for. We NEVER fail!"

After that I made sure to write up every late departure even without him telling me to. What we do as Airmen is important. How we do it is critical. We have to practice excellence in all we do, in everything we do, every time, all the time. We can't fail. **We will not fail!**

One of the most difficult things you will learn as a leader is that the cliché "It's lonely at the top" can be true. Kinda…. That's if you do it wrong. If you're working to be the top dog, you'll probably get there by yourself. And you'll find that that's a pretty lonely existence. So, bring someone with you! The more the merrier! If you expect to be an Exceptionally Powerful Lieutenant, you'll have to be care-full. You will also have to be careful not to get too close to your followers. "Wait a minute," you say, "first you tell me to take care of them, but now you're telling me to be careful of them. What's up with that?" Here's what I mean.

Who is accountable for the welfare of your followers and the accomplishment of the mission? You guessed it. It's you! Do you think that those two elements may come into conflict at any time? Surely, you've heard the old adage "Take care of your people, and they will take care of the mission." That's the truth. Taking care of your followers has a lot to do with nurturing an environment conducive to growth. Being care-full means holding your followers accountable so that they can grow and mature into good leaders themselves. How you do that is critical to your success as a leader. It often requires a different approach for every person you lead. When he was commander of USAFE, Gen John Jumper used a term that captures what I'm getting at. The term was *tough love*. I'm not sure who found it tough when it came to holding people accountable. Let me go back to my general for an example.

He was the best commander I ever worked with. His expectations were high, but I always knew what they were. When he first took command, he sent out printed pamphlets of what he expected from his wing. He was easy to follow because we all knew what he expected. I recall that he asked me (you don't really order Chiefs around) to do something. I agreed and went out to do it. Sometime later, I had to come back to him to tell him that I had failed (yes, yes, even Chiefs fail—but only once). I felt terrible that I hadn't been able to accomplish what he'd asked me to do. Evidently, he saw my frustration. After I explained to him what had happened—or hadn't happened—he looked me in the eye and said, "It won't happen again, will it, Chief?" Of course, my answer was, "No, General!" "Thanks!" he said. I'm sitting there, waiting to be admonished. Yell at me, man! Hit me! Do something! He just said, "Thank you" and dismissed me. That was the toughest admonishment I'd ever received. You see, he knew I didn't need any more than that. He knew I'd kick my own butt better than he could. Do you see why I'd go to war with that man any day, any place? Before I forget to mention it, his name was Gary Voellger, Maj Gen, USAF, Retired. I still love that man....

You have to know how to take care of your followers in a way that will help them grow and fulfill the mission. You have to be flexible in how you do that because your followers will respond differently to everything that happens to them in their lives. There is no "cookie-cutter" method for leading people. You have to show them you care, but at the same time you must separate yourself from them enough to be able to do what you need to do. That takes discipline. All of this sounds confusing, doesn't it?

Discipline is a word that we often use in a negative way. You can do that, but you'll get negative results. You'll think they're positive results because you may see immediate changes, but in the long term they will be negative. Let me explain. Here's one of several definitions of *discipline*: "punishment intended to correct or train." Now, is *punishment* a word that will get you positive results? If you answer yes, call me. You need help. We think of discipline as punishment, and it can certainly be that if you so choose. But the root word of *discipline* is *disciple*—"one who embraces and assists in spreading the teachings of another." Now we're getting into a positive tone, aren't we? Effective discipline comes from caring. Have you ever attended a gathering where promotions were announced? Remember the reactions of the followers when their new leader was introduced? "Ladies and gentlemen," the emcee announces, "I give you your new leader, Lt Bag O. Donuts!" The followers say nothing. The hush is deafening. Do you think there was much discipline or discipleship in that unit? I don't think so. On the other hand, consider this scenario: "Ladies and gentlemen, I give you your new leader, the Exceptionally Powerful Lieutenant Jones!" The place goes crazy. The troops hooah themselves hoarse! They are ready and willing to follow that lieutenant to war right now! Hooah, Lieutenant! What's the difference? The Exceptionally Powerful Lieutenant Jones takes care of her followers. She's care-full. She doesn't use punishment as discipline.

So how will you take care of your followers? If you'll recall, in Habit 3, "Look Up!", we talked about the power of saying "good morning." That's one of the little things that produces big results. Let me suggest four ways to show your followers that you're care-full.

First, tell them what you expect. Remember, I just told you I respected my general so much because I knew what he expected of me. Most leaders assume that their followers know what they expect. Often, they hold their followers accountable for doing what they assume their followers know is expected of them but never told them. Is that confusing or what? Try being in the middle of all that! I've been there, and it wasn't fun.

I remember, while teaching at the Air Force Academy, "leaders" complaining to me about the little, but important, things their cadets wouldn't do. "They don't stand up for me as I walk into the room, Chief. They should know to do that by now." "They think they can call me by my first name, Chief. (That's how she introduced herself to begin with.)" "They're so disrespectful, Chief." My reply to them was, "Who taught them otherwise? Who taught them that they should stand when a senior person enters the room? Who taught them that although a senior person may introduce themselves by first, then last, name means that it's okay to call them by their first name? (You know that's not so. Don't do that!) Who taught them how to be respectful?" The typical answer was, "Well, gee…I don't know, Chief." If you didn't teach them, who did? Don't assume someone else taught them. So don't assume that they know. Teach them yourself! Not by yelling at them, but by showing them. Because you will…. I've said it many times…. Your followers may not do what you say, but they'll do what you do. Sometimes "leaders" show their followers the wrong way.

I'm at the Air Force Academy. In the little Shoppette. My bride, Deb, works there, so I'm hanging out there, mostly bothering her. I hear someone speaking very loudly to a group of cadets. As I go over to investigate, I notice an Active Duty Officer yelling at a group of cadets for wearing flight suits. "Are you flying today?" he asks.

"No, Sir!" answer a few of them.

"Don't you know that you're supposed to wear the flight suit only on days that you're flying?" he's really upset.

"Yes, Sir!" they answer.

He goes on using some expletives that I won't use here and finally tells them to get out of his sight and go change uniforms. They take off very quickly. I know the officer, so I go up to him to try to calm him down, although I didn't actually do so. As he calms down, and I didn't do it to pick a fight, but I ask him, "Sir, are you flying?"

"No, Chief, why do you ask?" he replies.

"Because you're wearing a flight suit.," I remind him. He looks up at me, turns around and departs. I don't remember ever talking with him again. This point is in another lesson but let me tell you that they're always watching. *You're always setting an example, good or bad.* **Strive to set a good example.** Your followers will follow it.

The most powerful process I've found that will produce quantum positive results in dealing with expectations comes from Stephen Covey's book *The 7 Habits of Highly Effective People.* He calls it the win-win agreement, which includes five general areas: desired results (what you expect), resources (what you have to work with), guidelines (rules of engagement), accountability (who will be responsible for what), and consequences (what happens after success or failure). Key to that process is agreement. Both you and your follower agree to live up to the expectations in each area. When you both agree, that breeds discipline or discipleship for both parties. Another critical concept is to let your followers be creative. They have all kinds of ideas, so let them loose to share them with you.

The win-win agreement process takes time, but the investment will bring huge returns.

Second, know your followers. I didn't say go drink with them! Observe them. Listen to them. You know what that takes after having mastered Habit 2. Listen. Pay attention. Try to analyze their needs. Now, I'm not asking you to turn into a psychologist or anything like that. Just watch and learn. As the great philosopher Yogi Bera said, "You can observe a lot by just watching." What are they interested in? How can you help them grow? Think of them as winners and help them reach the finish line first. They will love you and follow you anywhere!

Here's an important lesson. This one's free: your followers, with very few exceptions, will value their families more than anything else. Learn as much as you can about their families and how you can help them invest more time with their loved ones. I know that may not be easy, but if you're interested—really care-full—they will appreciate and remember anything you do for them in that area of their lives. You see, taking care of work issues is easy compared to dealing with family issues. As the saying goes, "If Momma ain't happy, ain't nobody happy." If your followers are having family problems, they won't give you 100 percent at work. That's just how life is. So do what you can, even if it's not a lot, to help your followers take care of their families. If you do, they will go to war with you any day, any place.

I'd just sewn on Chief and had the honor of going back to where I started, March Air Force Base, near Riverside, California. I hadn't been there long when my commander, Major Denny Layendecker, summons me to his office.

"Chief," he says, "please, take my office door down."

"I'm sorry, Sir, (I didn't think I'd heard him correctly), did you say take your office door down?"

"Yes, Chief. Take my door down," he replies.

"Surely, Sir, but may I ask why?" I ask, bumfoozled.

"I want to have an open-door policy. I want the troops to know they can come in and talk with me whenever they want."

"Okay, Sir," I acquiesce.

Now, as I said, I'd just sewn on my Chief stripes. Chiefs don't do that type of work. They have Duty Sergeants who are training to be Chiefs one day. So, I find the Duty Sergeant and ask him to take the Major's door down. Yeah, I had to explain why. It was done.

About a month later I'm chatting with the boss and the topic of listening comes up. I ask him, "So, Sir, how's it going with your door down? Are people coming by to chat? You made that announcement in your Commander's Call."

"Chief," he looks at me, and in a bit of a saddened tone replies, "Not one person has come by."

"Hmmmm…," I murmur. "Your intent was to hear their thoughts, their problems, and learn about them, right, Sir?" He agrees. "Here's a suggestion. How about YOU go out there and hang out with THEM. Where they live and work. (The shop was on the first floor of a dorm. The Airmen lived on the floors above.)"

"I hadn't thought of that, Chief," he perks up. "I'll try it!"

A few days later he comes to my office and excitedly asks me, "Chief, did you know that Carmen is pregnant, and that Jerry's mom is coming to visit him next week, and that Smitty inherited a lot of money?" he goes on and on.

"Yes, Sir, I know all that." I reply.

"How did you know all that?" he asks, almost surprised.

"Because I do what you've been doing, Sir. I hang out with the troops trying to figure out what makes them tick so that when I make recommendations to you, they're based on what's important to accomplishing the mission and to taking care of our people. We can't be excellent until we know what that means and how we can do it."

"Thanks, Chief! By the way, can you put my door back up?" he asks.

"Can do easy, Sir," I answer. And then I went and found the Duty Sergeant.

Communication is key to success and even to having a good life. It requires listening, though, not just telling. Effective listening is done by listening with our hearts and eyes as well as with our ears. But we have to be there. We have to see and feel what's happening to truly understand. Listening is an active behavior. A leader can't wait or expect for her followers to come to her to talk, she has to find them, too!

Third, humble yourself. Yeah, you're the lieutenant. At least in your mind, you're the boss. Get over it! You're still accountable, but you're not going to succeed unless your followers become your disciples. If you're cocky and try to impress them with your authority, you'll quickly find how lonely it can be at the top. I can remember getting so much more done than I was capable of doing only because I could humble myself and ask young followers for help. "Man, the Chief needs *me*? I'm there for him! I've got his back!" they would say in their hearts. I'd then get out of the way and let them excel. I always gave them credit for their work when I got credit for the accomplishment.

I have little, if any, capacity for math. I can't stand numbers. I'm terrible at them. I've always been. One of several jobs I served in while I was on active duty in the United States Air Force was as Deputy Director for the Family Support Center at Ramstein Air Base in Germany. I didn't realize when I interviewed, and was selected, that part of my duties would include managing the budget. In other management positions I'd held I was responsible for maintaining the budget, but I always had numbers-people actually do the work. I was now responsible for a seven-figure budget. By myself! I can't stand numbers. Did I already tell you that? What to do?! What to do?!

No kidding, I had a sign appear before me. I was pondering what all the numbers in front of me on the computer screen and the green computer papers (You remember those?) meant when I just happened to look out my window. It was like a miracle! I noticed the word prominently displayed on the building across the street. It read "Finance!" It didn't include the exclamation point, that's just the excitement I felt when I had the epiphany that followed.

I can't stand numbers. You already know that. But I had to manage all of these numbers in front of me. Who likes working with numbers? THE FINANCE FOLKS! No kidding, I, immediately, gathered the green computer papers and walked across the street to the Finance building.

Upon entering the building, I noticed a young Airman hunched over her desk full of green computer papers, similar to the ones I had in my hands. I walked over to her, introduced myself, and asked her if she might be willing to help me make sense of all the numbers on my papers. She immediately lit up. She perked up, sat up straighter, and exclaimed, "You want ME to help you, Chief?" "Absolutely," I replied. "Leave your documents with me and I'll see what I can do," she encouraged me. I thanked her and went back to my office, to things I was good at doing.

A week later, that youngster brought me a folder of data all aligned and professionally bound. She explained everything that was in that folder and what all the numbers meant. I was incredibly grateful for her help. I made sure she knew so by the time she left my office.

Two months later, we had an inspection. Guess what the inspectors were most interested in? The budget! I'd done what that young lady had told me to do. The inspectors examined my program thoroughly. I received an Outstanding grade on my budget! The highest mark the inspectors could give!

Here's my point, again. If you're going to be a good, even GREAT, leader. You have to know yourself. You have to know, and hone, your strengths. As the great philosopher, Dirty Harry, once said, "A man's gotta know his limitations." That's true, but don't dwell on those. Know what they are but hone your strengths.

Hire someone to do the work you aren't capable of doing. And do what you're BEST at. Becoming an Exceptionally Powerful Lieutenant starts with knowing thyself. By the way, I helped that young Airman get promoted below the zone.

Never ask your followers to do more than you're willing to do. Lieutenant Hank Emerson was one of the best men I ever worked with. He was young and eager to be a good leader. He had one problem, though. He never asked his followers to do more than he was willing to do. Well, his problem was that he often did it for them! I remember the old Chiefs admonishing him to let the followers do what they were supposed to do, what they were hired to do. Bless his heart; Lieutenant Emerson cared so much that he wanted to do their work for them and spare them the trouble. At the beginning of this paragraph I said, "Never ask your followers to do more than you're willing to do." The lesson is, don't do it unless they need your help. If you're the only officer in the unit, your loneliness may try to overtake you. Your followers need to know that you're the leader. It's very difficult to lead from the trenches. Sometimes, you may need to do that, but not all the time. As a new leader, you don't have enough experience to do that effectively. Be willing to help and get your hands dirty if you're needed, but don't make that a habit. Immature followers will try to take advantage of it, and you don't need that bombarding you. I hope this makes sense. It's a paradox, I know, but it's true.

Humility is not the easiest virtue to embody. It takes a great deal of internal strength and a sense of purpose. If you're there to take care of your followers so that they can take care of the mission, you will excel because you will humble yourself to your purpose. If you're there to get promoted, watch out! Your followers will sense that in eight seconds and will make life hell for you. I believe it was Ken Blanchard who said, "Humble people don't think less of themselves; they think of themselves less."

Let me put it this way. You graduated from high school. For four or five years, you went to an academy or college where they taught you to be a "leader." You may have some type of technical training. You're all of 22 years old. You get to your first assignment where you're put in charge of an E-9 who's been in service 30 years! Do you think you might get a very quick lesson in humility from that E-9? Twenty-two years old! Man, I've got socks older than you! In fact, I'm wearing them right now. That's just the E-9. All of your other followers you're now in charge of have been to war more times than you've been to the bathroom! (Okay, that might be an exaggeration, but you know what I mean.) How are you going to lead those people? By following. By being humble. You'll learn more about that when you get to Habit 8. Know this, *humility is key to being an Exceptionally Powerful Lieutenant*.

I served at the United States Air Force Academy for more than 19 years. The best thing about serving there was watching our cadets grow into men and women of character. The second-best thing was working with incredible colleagues and teammates who made that happen.

Let me tell you about just one of them, who represents all of the USAFA Grads that I was honored to serve with.

This gentleman, and scholar, and warrior, and leader…. I could go on, but I won't, epitomizes all of those titles. He's a retired four-star general who served in many diverse command positions, retiring as Commander of U.S. Strategic Command in 2011. He's a 1976 distinguished graduate of U.S. Air Force Academy. Wow, you say, but that's not all! He's a three-time astronaut: STS-49 (Space Shuttle Endeavor), STS-59 (Space Shuttle Endeavor), and commanded STS-76 (Space Shuttle Atlantis)! Is that impressive? I would say so. But he's one of the most humble people I've ever met. By the way, he's General Kevin Chilton, Retired.

I invited his wife, who, too, is a person of great character, to join me during Basic Cadet Training, to help teach one of our character lessons. She was more than willing to do so, and, she asked, "Would you like me to invite my husband?" WOULD I?! ABSOLUTELY! So, she does, and he accepts.

If you've ever been to the Academy, you've quickly noted that there's never a parking spot open. Unless you show up at oh-dark-thirty like I used to. So, I set up reserved parking for the Chiltons. It's raining when they arrive. Mrs Chilton is driving and as she parks, she parks next to the reserved parking spot I'd designated for them. I'd set a "DV Parking" sign there, in the next spot. I'm watching this from under an overhang. As I notice she parked in the wrong spot, I'm about to go out to move the sign so that we don't take up two spots, that wouldn't be right. But before I get out there General Chilton gets out of the car and moves the sign himself. As I make it near him, I tell him that I'll move the sign. He tells me not to worry, he's got it.

So, here's a retired four-star general who served 34 and a half years, and has been in space three times, moving HIS DV parking sign in the rain! At first, I felt badly that I wasn't fast enough to take care of the sign myself, but he quickly made me feel better about it.

He, his wife, and I proceeded into the building to accomplish our mission of empowering the next generation of leaders as to what it takes to have good character. Well, he exemplified it! The lessons went very well, and the General stayed after the lesson to shake hands with all of the cadets who stood in line to meet this great warrior whom they all looked forward to emulating one day.

General and Mrs Chilton returned several times, to chat with our cadets, being hits every time. They didn't have to, but they wanted to, and they chose to. I was blessed to have worked with people like them. They call themselves USAFA Grads. I call them Leaders of Character!

Last, be consistent in your attitude and behavior. According to Don Shula and Ken Blanchard, "Consistency is not behaving the same way all the time: it is behaving the same way in similar circumstances." I've always preferred to work with leaders who were consistent. I could always count on how they would respond to what I did. Even those who were consistently negative were easier to work with than those who kept me guessing. My sense is that those who kept me guessing had no clear sense of purpose, nor did they practice humility. You should know by now that I would never advocate negative behavior of any type.

So, be consistently positive, and you will get consistently positive results. That's how life works.

That goes back to Habit 3 and having a positive attitude, doesn't it? You'll find that all of these habits intertwine to make you a total-person leader.

Chief Seattle of the Suquamish and Duwamish Native American tribes said in 1854 that "all things are connected. Whatever befalls the earth befalls the sons of the earth. Man did not weave the web of life: he is merely a strand in it. Whatever he does to the web, he does to himself." Those words tell me that what we do on a daily basis affects our world in profound ways, so the best thing we can do is understand that the people around us are more important than we are. The most exceptionally powerful people are humble. Humility is about being care-full putting others first. When we do, others will put us first. It's the circle of life. I used to tell newcomers to my unit that they had to follow two rules: (1) take care of the Chief, and (2) the Chief will take care of you. In truth, those rules worked backward, but they didn't have to know that. If you take care of your followers, they will take care of the mission, and they will take care of you. They *will* know how much you care by what you say, what you think, what you do, and how you do it—so be care-full!

Acknowledging a person's presence is critical to creating the type of relationship that will empower you and your followers to achieve greatness! Find each person you lead every day and ask him or her, sincerely, of course, how they're doing and how their family is. Ask if you can be of help. And be helpful if you can!

The great jazz singer, Ethel Waters said that "I am somebody cause God don't make no junk!" Each and every person has been born with specific talents that make them who they are. They may not tell you what those talents are, but you can find out fairly quickly if you pay attention. And when you do, acknowledge them for their greatness! We all like kudos every once in a while, so always thank them for serving with you! Acknowledge their presents!

A simple, yet effective, way to acknowledge your followers' presence is to find them and wish them a good morning, afternoon, evening, night, as appropriate. What you're telling them is that you care. As I've said, my friend, Simon Sinek, says that all people want is to be valued and valuable. I'll take it a step forward and say that every PERSON wants that and to be acknowledged for it. It's up to you as their leader to acknowledge each one daily. When you do, you're being care-full.

War Stories

My lovely bride of 45 wonderful and fulfilling years (the number increases each year, by the way) and I were married on 28 January 1978 (please feel free to send gifts on that day—or any other day, for that matter). By the end of February of that year, I was at Clark Air Base (AB) in the Philippines. We hadn't had time to get her passport, so she had to stay home in Charlotte, North Carolina, with her parents until she got her documents. I did all I could to expedite the process, but I was a young Staff Sergeant at the time and rather naïve about how things worked, so I wasn't very successful.

Hold on. Let me go back a little. In November of 1977, my mom contracted Guillain-Barré syndrome, a type of polio that almost killed her, hospitalizing her for many months. I found out about Mom's illness just after I'd gotten orders for reassignment to the Air Force Band of the Pacific at Clark AB. Because I had a report date of February 1978, I immediately asked that it be changed so I could be sure of her recovery before I went "across the pond," as we old guys say. My gaining unit commander denied the request because I was needed on a band tour since I was the only bassoonist.

Good warrior that I was, I left my mom in the hospital and my bride at home awaiting a passport and proceeded to my new duty station in the Philippines. After I made all kinds of calls and asked everyone I knew, including my commander, for help (who, by the way, never did), Mom got out of the hospital and Deb got her papers. I was relieved that Mom survived, although she has never been able to walk since, and excited that Deb could join me. I worked hard to make arrangements for her to fly to Clark, but I ran into a glitch. The only available flights would arrive during the time we would be on the tour for which I was so desperately needed. During the negotiations for changing my arrival date, the commander had fixed the band orchestrations so that another instrument covered my parts—as a precaution, I suppose. In other words, I really wasn't needed on that tour anymore. You can imagine how I felt having left Mom in the hospital so that I could do my duty, which, in the end, wasn't critical.

Under the circumstances, I asked to be left behind so that I could pick up my new bride at the airport and set up our household—you know, all the typical things that have to be done when you get to a new assignment. The answer, to my surprise, was no! I had to go on the tour. I went and didn't do much—but I was there. Deb had to wait another month to join me. Again, can you imagine how I felt? Do you think I was a gung-ho warrior at that time?

I'm not bitter now. Okay, maybe a little. But I was then. Why? My commander was not care-full. I never got to talk with him. The E-9 we had in the unit was the only person who gave me an audience—but he didn't care either. I never understood why they had made their decisions, and they never tried to explain them to me. I suppose they didn't have to, but they certainly would have gotten more out of me if they had.

One of the good things about military life is that everything changes when a new commander comes in, every two years or so. Or does it? The commander and E-9 who had no idea how to be care-full were reassigned, and two new people took command. "Hooah!" I thought. I'll admit they were more care-full than the previous regime, but something happened that I hope you will never allow, although I see it in units to this day. You have to stop it! Let me explain.

As I said earlier, I was a naïve Staff Sergeant just doing my job. But I was a hard worker, so I deserved to get promoted. And I got promoted, although I almost didn't know it. Here's what happened. Deb and I were at the park in the housing compound where we lived. For some reason, ironically, I hadn't gone on this particular band tour. The band had just returned, and at the park I happened to see the E-8 (the one who had replaced the previous E-9) since he lived in our housing area. After the usual salutations he said, "By the way, congratulations!" "Congratulations? On what?" I asked. "On your promotion," he replied. "What promotion?" I asked, confused. "To Tech Sergeant. Didn't anyone tell you?" he asked. Obviously, the answer was no. No one had told me. Deb and I should have been jumping for joy, I suppose, since it *was* good news. We were more confused than elated, though. That was not exactly the best way to learn about a promotion.

There are two lessons here. First, as I said before, your followers' families will almost always be the highest priority in their lives. If you help them take care of their families, they will do anything you ask of them and more. I certainly would have been more productive at that Philippines assignment if I'd trusted my leaders to take care of my family. In truth, though, you would have had to shoot *me* to go to war voluntarily with any of those men.

Second, you'll have to do some unpleasant things in your duty as a commander—too many of them. One of the few *good* things you'll do is promote your followers. That should be a momentous event. When your followers get promoted, find them—no matter where they are—and present them their new stripes. If they have family and you can include them, do it! Promotion is the acknowledgement of value. We all want to be valued. I know you've heard that before. The small things you do every day to express how you value your followers are incredibly powerful. But a promotion is the institution's formal way of expressing its respect and trust in a warrior. It should be an unforgettable event—a positive, unforgettable event. I guarantee you that we all remember every one of our promotions. And we remember who was in command. If you're care-full, you'll spend less time being careful because you'll have a team of experienced, knowledgeable, caring warriors around you doing all they can to help you become an Exceptionally Powerful Lieutenant!

Here's another story, but this one is about how to ensure people are respected and valued even when you don't know them. CMSgt (deceased) Leigh Steiger was the First Sergeant for the United States Air Force Band when I met him. He was a truly great man. I was a Master Sergeant. I was TDY to his unit the day that the line numbers for selection to the rank of Senior Master Sergeant were announced. I'd been called by my Chief, Dave Griffith, to tell me I'd been selected. I was grateful for that. As I walked into the Air Force Band's building, I found Chief Steiger in his office. I introduced myself and we started chit chatting. All of a sudden there was a huge commotion right outside his door that led into a large conference room. As we both went out to see what was going on, we realized that the commander and a bunch of other folks were out in the big room celebrating. He'd just announced who, in the unit, had been selected for promotion. There was a lot a cheering going on. Once we applauded and hooped and hollered for the selectees, Chief and I went back into his office. In as humble a way as I could muster, I told the Chief that I, too, had been selected. He beamed, got out from behind his desk, and shook my hand as he congratulated me. He then took me out to the big room where there were still people partying and asked them to quiet down, then proceeded to tell the group, who didn't know who I was, but it really didn't matter, that I had been selected, too. They all applauded and congratulated me. Chief didn't have to do that. His congratulations would have been enough. But he valued me, whoever I was, as well as the importance of the occasion, enough to make it public. I'll never forget that day, how I felt, nor My Brother Chief, Leigh Steiger. Here's a toast….

As I previously mentioned, I served in several Air Force bands (most, unfortunately, don't exist anymore). While serving with the Air Force Band of the Golden West, stationed at March AFB in California, we had many opportunities to play for President Reagan at various high-level events. Evidently, he liked us because after he retired to the Los Angeles area, he called us to come visit him so that he could thank us for what we'd done for him while he was in office. Now, think on that a little bit—a former President of the United States summoning members of an Air Force Band to his office to thank them! That in itself is being care-full. He didn't have to do it. But he did!

We took advantage of the invitation and went to his office one bright, sunny day. (It's always bright and sunny in California, by the way.) What an exciting time! The president invested more than an hour with us, joking around and telling stories. What a great storyteller! We all had our pictures taken with him. (I have mine proudly displayed in my office if you'd like to come by and see it.) Apparently, his next appointment was some golf tournament. He was dressed casually since he had to go to the course right after talking with us. I was the second-to-last person to be photographed with him. My commander was last. I'd trained him. As we finished taking the photos, one of the President's aides came in and reminded him that he needed to hurry to get to his next commitment. The President acknowledged the remark but finished the photos. Again, he could have apologized and rushed off with his aide, but he stayed with us until we were all done.

I'd brought my copy of *An American Life*, the president's autobiography, hoping to have it autographed. We hadn't had time for that, however, so as we were being led out of his office, I asked another aide at a desk near the door if she would please ask the President to autograph my book and I would gladly leave enough money with her to cover the postage to mail it to me. Although President Reagan was being escorted out another door, he stopped, having overheard my conversation. He told the aide with him to get my book, signed it, had his aide return it to me, and then went on his way. Did he have to do that? Of course not. But he did! He cared enough to stop what he was doing to take care of someone he didn't know very well but respected. He even remembered my name and, most importantly, my rank. He signed it…

> To Chief Bob Vásquez—With Best Wishes
> Ronald Reagan
> May 17 '93

You're welcome to come see my book as well. I won't let you touch it, but you can see it.

Many people loved President Reagan. Even more respected him. Surely, history will remember what he did for our country, but the people he touched will always remember him for how he treated them. He was care-full. I would have gone to war with him any time, any place. God bless you, Mr. President!

Starting Points

• **What do I know about my followers' families?** Are they in good health, and if not, how can I help them? Do they have any recurring issues that I can help them work through?

• **Do my followers know what I expect of them?** Do they understand what those expectations really are and why I expect them? Telling them doesn't ensure that they understand. You're their boss, their leader. If you ask them if they understand what you just said, they will nod their heads and say yes even if they have no idea what you actually meant. Ensure that they understand the expectations so that you can measure whether they understand them and then meet them. The best way to teach is to show, so if you can find someone who is already meeting your expectations, show the new person what that looks like. You may even have the veteran show and teach the rookie. Peer teaching is powerful.

• **Do I practice humility?** As I often say, the only person who can measure whether or not you're being humble is you. But others, pretty much everyone, will have a perception of your humility. You might ask a trusted mentor whether or not you're seen as humble. You may adjust your behavior to meet others' perception. That's up to you. But you're the only one who will truly know. Consider that as you do what you do every day.

• **Am I taking care of my followers, or am I just saying it?** Many "leaders" don't practice self-assessment because they're afraid of the results. One way to develop humility to through self-assessment. It begins with this question. And there's actual evidence of the answer. At least weekly, take some time to consider whether or not you've taken care of your followers and how. Be specific. Capture the results. If what you're doing is working, keep doing it. If not, do something else or differently. It will be easy for you to get caught up in the thick of thin things, as Dr Stephen Covey taught me. Make the time for self-assessment and be honest with yourself. Note that Exceptionally Powerful Lieutenants will probably rate themselves low in this assessment. That's being humble. But if you're objective about how you're doing, there's always room for improvement. No one is perfect, but you can be exceptional!

Words of Wisdom

The greatest carver does the least cutting.
Tao Te Ching

It is our choices, that show what we truly are, far more than our abilities.
J K Rowling

People will sit up and take notice of you if you will sit up and take notice of what makes them sit up and take notice.
Frank Romer

We shall never know all the good that a simple smile can do.
Mother Teresa

A desk is a dangerous place from which to watch the world.
John LeCarré

The shortest distance between two people
is a smile.
Victor Borge

We can do no great things—only small things
with great love.
Mother Teresa

No act of kindness, no matter how small, is
ever wasted.
Aesop

Our values are always in our hands.
bob vásquez

Habit 5

Sharpen the Sword!
Take Care of Yourself First

A lieutenant is a leader. A lieutenant leads warriors. A lieutenant *is* a warrior! We use the term *warrior* extensively in the military, but what exactly *is* a warrior? A typical dictionary definition of the word is "one engaged or experienced in battle." I'm not sure how much *experience* you have in battle, but you're certainly *engaged* in one. Before you start hooahing, let me ask you this: where is the greatest battle in your life? Is it not inside you? Isn't making yourself do what you should do a continuous battle? If it's not, you're really the exception. We'll have you canonized as soon as you die. The biggest battle most of us will ever fight is the internal one. Joseph Campbell, American writer on mythology and comparative religion, talked about the "dragon within" when he referred to a person's daily internal struggles. What would it take to slay that dragon within? Wouldn't it help to have a sharp sword? Where do you find that weapon? The sword is within you. I'll soon tell you how to keep it sharpened.

First, let me ask you a few more questions. Have you ever known someone who seemed tired all the time? Maybe that person was someone you avoided because he or she was always moody and hard to get along with. Have you ever heard of someone who was on the fast track to success but burned out instead? A better question might be, are *you* that someone? If you have ever suffered burnout, you need to develop and hone the habit I discuss here—and you need to start now!

In Habit 4 I told you about the value of selflessness: the importance of serving others by being care-full. So, you might think I'm being duplicitous when I tell you to make a habit of taking care of yourself first. This notion may go against every leadership lesson you've ever learned, but it's the truth. I would never lie to you. (Okay, unless it was for your own good.)

Here's the deal. Although serving others is key to being an effective leader, truthfully, what good are you if you're not around? How effective are you when you're ill? Can you "hang" with your followers when you're worn out? Do they perceive you as an effective leader when you're not in tune with what's happening in your environment or your specialty? When you're having a bad day, does it affect those around you? How inspiring can you be when you're not inspired yourself?

Surely, you've taken a flight on a commercial airline. During the safety briefing, the flight attendant will get to the part about the little yellow oxygen cup coming down in front of you in case the plane loses altitude quickly. The attendant's message is to don the mask *first*, breathe normally THEN take care of your kids. Breathe normally. Yeah, right! If that little yellow cup comes down in front of me, my first reaction will be to breathe as fast as I can—it's somehow associated with Abraham Maslow's hierarchy of needs. Survival is the first need!

Now, assuming I even remember that my wife and kids are with me on that flight, I'll certainly make sure that they have their masks on, too. My gut tells me that this is an especially difficult concept for moms. Dads are from the old school: "I brought you into this life, and I'll take you out!" Moms will sacrifice *everything* for their children. Men think, "Hey, we'll just have more." (Okay, not really. Okay, maybe.) Anyway, the idea of taking care of yourself first is credible in this situation. It probably won't take you long to don the mask. If that's the only way to breathe, you'll do it with incredible speed. Rather than encouraging us to breathe normally, I think that the attendant really wants to keep us from panicking. That's one of the vital things you learn in CPR classes— don't panic. Emotional minds cause panic. Once emotion takes over, you can't think logically, and the fight-or-flight instinct assumes control—not a good thing. If you take care of yourself and don't panic, doesn't it make sense that you'll be able to assess what's going on a little bit better and make the right decisions? Then you'll be able to take care of others. I think so. Serving others begins with serving yourself.

Your life includes four basic areas: physical, mental, emotional, and spiritual. If you're going to serve others, you have to be healthy in those areas. Again, take care of yourself first, and then you can serve others. Let me break it down for you.

First, let's talk about *physical fitness.* How important is it? Of course, you know you have to take care of your body as long as you're in it. When you're not in it anymore, don't worry about it. For decades, Americans have been on a health kick. Still, we have ungodly numbers of people dying from heart attacks and strokes. The number of obese people increases continuously. Something's not right there. Perhaps it's the New Year Resolution Syndrome. Have you ever noticed that you can hardly get into the gym during the months of January, February, and March? By April you can walk in any time because it's almost empty. My biggest problem with maintaining physical fitness isn't so much that I don't have a program; it's that I won't stick with it consistently.

Is physical fitness important to taking care of yourself? Of course, it is. What does it consist of? Most of us quickly develop the vision of looking like Jason Mamoa or Jennifer Lopez. That's okay if you have the time and can make the commitment to it. But do you know what it takes to look like that? A lot of work! Most of us aren't willing to make that investment, but we still want those results. It won't happen! But you *can* maintain your health and even retain your girlish figure if you work at it consistently.

For years, physical fitness experts have told us that ***maintaining our hearts in good working condition requires only 20 minutes of aerobic exercise a day, three times a week.*** Twenty minutes, three times a week. If my ciphering is correct, that's one hour—per week. Do you know how many hours there are in a week? One hundred sixty-eight. (Trust me, I did the math.) Is it worth investing 1/168th of your time to increase your life span? If it increased by only one day, would it not be worth it? "Well, yeah, but I don't have time, Chief," you rationalize. "But I can't!" "But my shoes are too old!" Which "but" do you normally prefer? I already told you what you produce when you rationalize.... Rational. Lies.

Now, get off your "buts." I'm not talking about lifting hundreds of pounds at the gym, but if that's what cranks your motor, great—do it! All I'm talking about, at a minimum, is taking a brisk walk for 20 minutes. Imagine! You're behind that computer you love, replying to every single e-mail for hours on end. Wouldn't it be nice to get up and take a walk for 20 minutes? If you did that twice a day, wouldn't you feel better and possibly increase your life span? Simple stuff! All you need is a place to do it (outdoors is great, but the mall will work, too) and a pair of walking shoes. Part of our problem, I think, is that we want to look cool while we're exercising, and that becomes the priority. You don't have to look cool to be healthy. That will come later when you *are* healthy!

The other part of physical fitness has to do with diet. I'll tell you now, this is the most difficult area of *my* life. I'd rather work out an hour a day, every day, than to give up the foods I like. I suppose it's better to strike a balance, whatever that is. Evidently, a balanced diet is *not* a Big Mac in each hand! Again, you know the importance of not eating too much fat, eating whole grains, and drinking plenty of water. But I'm not an expert by any means. Get a good book that will help you eat sensibly—and not too much. I can't tell you how many times we Air Force Bandsmen would go to a performance site after we'd had dinner only to find food set up for us. Although we'd just eaten, this food was *free*, man! Kids in Africa are starving, so how could we possibly waste free food? Luckily, most of our concerts were of a lighter vein. (Hey, I'm trying to keep you awake here!) The point is that **diet is critical to physical fitness.** Learn that before you have a heart attack, and you'll be much better off. What's almost ironic is that you already know this! But do you apply it?

Proper rest is critical to maintaining your physical fitness. I remember a TV commercial that professed that "the best never rest!" Yeah, well, they die young. The body is not constructed to work 24/7. We often think we can work that hard, but it ain't so! It needs to replenish its resources. Myriad studies conclude that if a person doesn't get enough rest, he or she will break down psychologically as well as physically. I know there's not enough time to do all you're asked to do, but if you don't rest, you'll eventually do nothing for a lot longer. Get plenty of rest, and make sure your followers do the same—or else! The "or else" will manifest itself in illness, weakness, and fatigue, all of which can compromise the mission. You know it's true.

Second, let's consider *the mental area of your life*. Let me add that maintaining physical fitness has an effect on our mental well-being. I recall that for most of my military life I was told, "We don't pay you to think; we pay you to do." Maybe, 100 years ago (20 in military years), that might have been the case. Now, with technology changing our world every day, you and your followers *have* to think. ***Our military suffers in two areas: not thinking and, when we do, what we think about.***

Imagine walking down the hallway of your duty section only to find a young follower sitting there, doing nothing. I mean, he's just sitting there! *Doing nothing*! Ha! I'm a lieutenant! I'll fix that! "Young man, if you have nothing to do, let me find you something!" you say. Hey, I've been there, done that! You feel much better, don't you? Well, maybe you shouldn't. Is it possible that that young person was *thinking*? Is it possible that he or she was *this* close to coming up with a better way to perform your mission? Is it possible? You'll never know now, will you? How is it that we improve anything in our lives? Is it not through thinking? Someone thinks of a better way. Usually, it happens on the fly. Imagine if you and your followers made time to think of new ways to do things!

Regarding what you think about, I tell you, if you don't get your followers to sit down and think about something specific, they'll think about everything else. Have you ever asked one of your followers or yourself, "What *were* you thinking?" It's usually after they—or you—do something wrong that you ask that question. You *do* pay your followers to think! And you have to train them to do so!

So how do you exercise the mind? There are plenty of ways, but if you can't think of any, I have some suggestions: reading, for one. *Leaders are readers.* The problem with reading is that it takes time. None of us has time to read. We have to make time! You make time for watching *Survivor*, don't you? As with physical fitness, I'm not suggesting that you read *War and Peace*. Make 20 minutes available each day to read books about your favorite subjects, hobbies, or work. You might find some peace of mind. You might even learn something. The more you do it, the better you'll get at it. *Read and learn. As you do it, your followers will see you and know that you value learning*.

Hone your skills. More on that in the next habit but suffice it to say that you can always find interesting classes to attend. Oh, I know, you're always on the road, the ops tempo is a killer, and the dog eats your homework. What do you produce when you rationalize? You know what. Get over it! As Hannibal said to his followers, "We shall find a way or make one!" (Or something like that. Hey, I wasn't there!) We have the means to do most anything nowadays. Technology connects us with people around the world. If you want to learn more about your profession, you can find a way to do so. **Continue to learn.**

Listening is an inexpensive way of exercising the mind. I already taught you how. Now practice it. Knowing but not doing is not knowing. Take what you read about in Habit 2 and practice it daily. You may be amazed at what you can learn when you listen. Take that learning and use it to accomplish your mission of either doing your job better or taking care of your followers better. You'll improve your mental health as well as the health of your followers. I'm going to sound duplicitous again, but when you take care of your followers, they'll take care of you! **Don't waste your brain. Feed it daily.**

Third, you must hone your emotional life if you're going to lead anyone. The first person you lead is you. I barely recall, about two centuries ago, the movie *The Greatest*, about Muhammad Ali—not the new version starring Will Smith, but the "real" one. The lyrics of the theme song, "The Greatest Love of All," later recorded by Whitney Houston, referred to the greatest love of all as the love for oneself. At the time I thought, "How selfish!" As I matured (not aged), I came to realize that loving oneself *is* the first step to emotional stability. One cannot give what one does not have. (Wow, I hope I just made that up!) If you don't love yourself, how can you love anyone else? It's impossible! Life is about relationships. We often work on developing relationships with others. We also have to maintain a relationship with ourselves.

I hope you never have the experience of being around someone who's suicidal. When you become a commander, you'll probably have to deal with that situation in some way. Suicidal people have absolutely no love for themselves, and they will bring everyone around them down emotionally. You've been to those suicide prevention briefings every year. But they never get to the real cause: the lack of self-love. The only way to help those who don't love themselves is to love them until they do. That's not easy, but it's the truth.

Now, don't get me wrong. I'm not suggesting going to extremes, such as becoming narcissistic. But the ego must be strong, not big, for you to do what you have to do, and that comes from self-confidence, self-esteem, and self-love. Then you can share that love. In fact, I believe it happens almost naturally.

Okay, let's talk about the *L* word. I know it's touchy-feely, but the most important things in your life are touchy-feely, aren't they? What's most important to you? If you don't say family, you're probably reading this around your buddies. I'll wait for you to be alone, and then you'll say family. What keeps a family together? It's love, man! **John and Paul were right: "All you need is love."** Keep in mind there's a huge difference between love and lust. To put it in simple terms, love is unconditional giving. Lust is selfish wanting. Easy enough?

Every relationship can be measured by the amount of love shared. Think about it. (Yes, now I'm asking you to practice what I preach.) Your relationship with another person is strong because you're willing to give him or her all you can—all you have. You're willing to give without the condition that he or she give back to you. That's love! And there's incredible power in that! As I write this, my heart is saddened by the overwhelming number of our warriors who were killed in our last war. Having been a warrior all of my adult life, I know that a Soldier, Airman, Sailor, Marine, Guardian, or Coast Guardsman can do what is expected only through the strength gained from love of country, family, and fellow warriors. What can a person who puts his or her life on the line expect in return? Money? That's silly, isn't it? By now you should realize that you're in this, not for the money or prestige, but because you love what you do and who you do it for. It's about relationships. It's about love!

Last of all, *the most important part of your life is spiritual*. Interestingly, when you've taken care of the physical, mental, and emotional areas, you'll find a sense of peace that empowers you to think beyond yourself. Maslow called it self-transcendence. It goes beyond self-actualization. I call it spiritual.

Often, people misconstrue spirituality with religion. It's not necessarily the same. If religion is the way by which you develop your spiritual life, then hallelujah! If it's not, don't worry about it. To me, spirituality is about opening your mind and your heart to appreciate all that abounds around you.

I live in Colorado, near Pikes Peak. I served at the Air Force Academy for 19 years. Every morning I drove into work, I'd drive toward the west and get a glimpse of a giant postcard. I'd see what Katharine Lee Bates saw when she wrote the lyrics to "America the Beautiful"—from a different angle, of course. The mountains really were purple, and they were majestic! Paradoxically, one of the most effective ways to get to the spiritual is to appreciate the natural. Go outside, whether in the sunshine or the snow, and listen and observe. Hear the birds and the wind. Feel the breeze. Nature will astound you with its beauty if you pay attention.

Another way to develop your spirituality is by thinking (there I go again, using the *T* word) about the value of the people with whom you live and work. Time has changed the risks of living in this world. There's a possibility—not a probability, I hope—that all of those people with whom you live and work and, possibly, those you love could get killed as you read this passage. God forbid, but the possibility does exist. We are not as safe as we used to be. Imagine—again, God forbid—that your friends or family were all killed. How would you feel? Would that not put you in a spiritual frame of mind?

In Habit 4, Be Care-full, I shared a story about my genius boss and his idea of having an open-door policy. He taught me a lot about caring. Here's another story about how to develop a caring attitude.

I'm sitting at my desk at Ramstein Air Base in Germany. I'm serving with the Air Force Band in Europe with a genius of a man. The building we worked in wasn't soundproof so you could hear the music being rehearsed down the hallway. The concert band is being rehearsed by our commander. I hear the band starting and stopping and someone speaking loudly in between. It continues several times until the music stops abruptly. All of a sudden, I hear a door slam and shortly afterward I hear my boss marching down the hallway. As he passes my office, the door of which is open, he says strongly, "Chief, my office!" As a good follower, which every effective leader is, I follow him, curious as to what just happened.

Man, he's pissed! As soon as I close his office door, he begins ranting and raving. Raah, raah, raah, he goes on. I let him finish before I ask him, "Sir, what just happened?" not being able to understand his raving and ranting. He was upset with the band. At first it seemed because they weren't playing to his expectation, but, really, it was because they weren't playing to their capability. They could play much better, they just weren't doing so and he was frustrated because he couldn't make them flourish.

"Sir," I said to him in my calm voice. "What if we were all on our busses, heading to a gig, and something terrible happened? How would you feel?" "Well, Chief, I would feel terrible. These are great musicians and great people. I don't want them to be hurt in any way." I went on, "You realize that we're all in danger every day. Especially when we get on those busses that advertise that we're American Airmen, don't you? There's always a risk that some of us, or all of us could get killed." He sat looking at me for few minutes, pondering my thoughts, and then he said, "Chief, please get the band back together in the rehearsal hall." I did as he requested.

A few minutes later, the boss is back in the rehearsal hall, and I'm back at my desk. I soon hear some of the most incredibly beautiful and moving music I've ever heard. I'm told that the boss went back in, looked them all in the eye, and led them to be their best selves for a few moments.

Surely, we don't know what we'd do if a catastrophe like that befell us. In truth, we do know what we should do to express our appreciation and love for those with whom we share our lives. We often choose not to because we don't exercise our spirit. You can change that. Today. Right now!

I hope you'll practice this habit daily. It will affect you, and it will affect those with whom you live. If you choose to do nothing else I've commended to you, work on this habit. It will increase your life span, and it will give you peace. Take care of yourself so that you can take care of others who, in turn, will take care of you. *Developing and maintaining these four basic areas of your life is critical to your ability to lead warriors.* You have to have a sharp weapon. Denise Austin, the fitness guru, said it best: "Health is wealth!" Get rich! *Take care of yourself first!*

We all have different experiences that we learn from that help us Sharpen the Sword. Here's a great perspective from one of my protégés, Master Sergeant Joshua White.

Sharpen The Sword!

Physical - Mental - Emotional - Spiritual

(Bonus: Gratitude)

As we roll into yet another COVID spike, I'm reminded now more than ever of the importance of keeping your sword sharp; being ready for battle, ready for life. When the stuff hits the fan and we are going to war, there is no time to sharpen your sword...you need to do that *before* battle. Once you're in the fight, you'd better hope that that bad boy is sharp and ready! With that being said, I will share a war story for each pillar. This won't be the normal advice you've seen on staying resilient...Instead, I'm going to show you what happens when you don't.

Physical

Earlier in my career I contracted a heart virus from the flu mist vaccine that put me in the hospital for two weeks. Part of my diagnosis was to apply a cardiac catheter, to string a small camera through the artery in my groin, up to my chest, so the docs could see real-time how my heart looked. Obviously, this whole event shook me to my core, but that's another story. The aftermath is what I want to discuss. Upon my return to duty, I was exempt for taking a PT test for nearly a year. I stopped exercising and forgot about it. As the months went by a thought occasionally popped into my mind, "Oh no, when was that date again?" The thought of my pending test would fill me to the brim with anxiety, so I pushed it out of my head. I chose to not keep my sword sharp.

One day my flight chief called me to his office and informed me that my PT test was due this month. I felt my stomach drop. "Had it already been a year? How could this happen? Why didn't I prepare?" I started to panic thinking about what my career would look like when I bombed this test. Later in the day my flight chief informed me that my test date would be in a week. Could I train for a week and pass? I wasn't sure but tried to stay positive. Right before I left for the day, my flight chief approached me one last time. "Your test is first thing tomorrow morning." WHAT!? I was doomed. I'll never forget how I felt that morning as I was pushing myself to the brink of death to pass that thing. My legs felt like cement, my lungs were on fire, and my heart felt like it was on the brink of exploding. My anxiety level was through the roof. During the entire run, all I could think about was the consequences I would soon face for failing the test. I was running for my life (only someone in the military understands this feeling), but I somehow passed by a few seconds. I vowed that day that I would never put myself through that kind of mental anguish again. Putting my problems off was not a good strategy. After 17 years in the service, I've never failed a PT test.

Takeaway — Avoidance ALWAYS makes things astronomically worse. Stay accountable to yourself.

Mental

When I read the word "mental" *a lot* comes to mind. Things like going to school, learning something new, thought-provoking group discussions, or reading. Most of us are already aware of those and understand the value in them. After all, leaders are readers! But what about your *mindset*? The way you see yourself and the world. My first ten years in the Air Force I struggled with a fear of failure. It became easy to just avoid getting out of my comfort zone altogether. What I've learned is that leaving my comfort zone had a drastic impact on my ability to elevate thoughts and break mindset barriers. Barriers I didn't even realize I had.

For me, my mindset changed in only a few short years, and it all began with coordinating and running a Group Commander's Call. Leadership was looking for volunteers, and at this point I would usually shy away. With the encouragement of a senior mentor (who really didn't give me a choice!) my name was submitted, and I was selected. What I didn't realize was that the simple act of taking one step toward something I was terrified of would change the trajectory of my life. Little by little, I showed myself what I was capable of.

After this successful event I took on a Wing 5K. Then I was an Additional Duty First Sergeant, a Group Executive Officer, and, finally, I was selected to run the Base Honor Guard, which was a life-changing experience and something I previously could *never* have seen myself being able to perform. Leadership books, discussions, and education showed me what success looked like, I had the roadmap. It wasn't until I stepped out of my comfort zone that I was able to benefit from it. In only two short years my entire life had changed.

Takeaway — Sharpening your mental sword is useless without application. A fear of failure prevents action and leads to missed opportunities, which prevents growth. My dad used to tell me "Fear is the sand in the machinery of life." And making mistakes is part of the process. Understand it won't always look pretty, and that's ok, so start small and build momentum. The success will come faster than you think!

Emotional

If there's one pillar the military neglects the most, it's this one. The stigma surrounding mental health in the military has been a part of our military culture for decades. PTSD, trauma, abuse, bullying—all these unfortunate events can come back to haunt us. When we're younger I don't think we notice as much because it's our "norm" and we're not yet depending on our skillsets and motivation to be self-reliant. Once we're older we start to reflect on our past and share stories with others. Through these stories we learn that our "norm" actually wasn't all that normal. Furthermore, the new set of challenges and pressure the military imposes on us can have us react in unpredictable ways.

Your past trauma is like a rock thrown in a still lake. The ripples become wider and reach in all directions as time goes by. Things you should have paid attention to start to fester and manifest in ways that are detrimental to your overall health as well as your mental health. Drugs, alcohol, food, toxic relationships—all these and more can be physical manifestations of trying to cope.

I convinced myself that I didn't have a mental health or coping problem and so I certainly didn't need to talk to anyone about it. Instead, I would go out on the weekends and get hammered. I felt free, energized, and confident. The anxiety and stress were gone. What started as weekends of fun turned into a nightly routine of isolating, gaming, and drinking. My health declined and, over time, things got much darker for me. I started to question my place in the military and even the world and I felt like all my dreams and aspirations were a lie. Truth was, I hated myself. At that time, I was in Maintenance on the flight line, never promoted, and just coasted. I wasn't a good Maintainer and, given my competitive nature, that really ate me up inside; my insecurities grew to new heights. Here's the thing: for years I never did anything about those feelings and, consequently, didn't take care of myself.

So, what does avoiding therapy (when you need it) look like? My emotional health plummeted, and I had no mental fortitude, much less a sharpened sword. Negative feedback or confrontation—any little thing—would set me off. I interpreted virtually everything as an attack, and I couldn't handle it. At about six years in the service, I was at my breaking point. Luckily, I was able to get a month off work and my parents flew out to see me.

During that month I was able to relax and not think of all the things that stressed me out. Although this was a nice break from the noise in my head, I still didn't take the time to truly sharpen my sword. On the last night before returning to work I finally broke. I had a full-blown panic attack. Crying, and not able to catch my breath, I felt my emotions escalating and I lost all control of my thoughts. Fearing for my safety, I called my dad in the middle of the night. He answered on the third attempt and was able to calm me down. Had he not answered, I'm really not sure what would have happened. But the next day I put on a smile and pressed on, no one knew about this "episode" or my current state of mind.

Take away — You can't out-run your demons. They are always with you, waiting to strike when you're most vulnerable. You have to learn to stay ahead of them by talking to someone, asking for help, being honest with yourself, and finding the strength to talk about your issues out loud. Get ahead of it, keep the sword sharp.

Spiritual

Spirituality is always a touchy subject because of the many different beliefs people serving in the military have today. Truthfully, I believe in Christianity. However, I'm not one to frequently pray or attend church services. Those things are certainly "spiritual" to me, but I feel there's more to spirituality than just those things. For me, being "spiritual" is to have deep, meaningful connections with others. It's being vulnerable and operating from a place of love. It's having a strong sense of purpose, meaningful relationships, and a deep understanding of your values. The opposite is being selfish, unaware, withholding information, gossiping, isolating, attention seeking, and not celebrating others (especially your peers). Your ego is guiding your emotions.

In 2019 I became one of the 12 Outstanding Airmen of The Year (OAY) for Air Force Global Strike Command as the Honor Guard Program Manager. An OAY is an incredible achievement that most Airmen never reach. It's being one of only 72 Airmen, out of roughly 328,000, selected for that respective year. Before that accomplishment, I hadn't won a darn thing. Throughout most of my career I'd struggled to maintain my spiritual pillar but was filled with ego and insecurity. I didn't have love in my heart, though I did experience fear. But after the OAY experience of 2019, I felt validated—like I cracked the code for success.

I met all the leaders who ran the Air Force. I became part of their network, and I finally felt like I mattered. In my mind, this was my new life and there was no going back to the old Josh. After the six-month high of winning the award, things died down and life went back to normal. Public Affairs wasn't interviewing me, I wasn't in the base paper, I wasn't going on exciting TDYs, and my special duty (Honor Guard) was ending. Soon, I wouldn't have that super elite and high-vis mission, which was a problem because by this point it was my identity.

It was around this time that COVID hit, and we were on lock-down. During that first month, I had a lot of time to stew in my soon-to-be new way of life and perceived irrelevancy. As a result, I went on full attack mode knowing my ego was fighting for its life to stay relevant. After my 12 OAY experience I felt that I would never need another award again to feel validated and fulfilled. During the lockdown, this feeling changed, and I decided to submit myself for the OAY award once more. I was told it was highly unlikely for anyone to win two years in a row at that level but the feeling of being invisible was overbearing and I went forward with it, along with many other awards.

I poured hours into the award packages and even reached out to past winners across the Air Force and received copies of their submissions to better guide me. I was confident that I had the best packages, however, out of four submissions not one of them made it past the first stage. I probably looked like a jerk, but even worse I felt like a jerk. The truth is that my spirituality was wrapped up in a title versus myself, God, or family. I was a husband, father to two children, and, by all measure, successful in so many other areas of my life—this is what I now keep close to my heart.

Takeaway — Your life's purpose, values, and relationships should have a foundation rooted in what you actually love the most. A title, position, or award is temporary and will be old news much sooner than you realize. The people in your life, the places you've been, the lives you've changed—those are joy, those are forever.

Gratitude

Gratitude is something I've struggled with my entire life. I understand the concept but wasn't sure how to exercise it. Was it simply saying thank you? Was it meditating on what you're appreciative of? I thought maybe I was just wired differently. The truth is that everybody exercises gratitude differently and I just didn't have the emotional intelligence to know if I was exercising mine properly. When my emotional IQ was much higher, I became more vulnerable. Vulnerability is something I now consider a superpower, but it's something we have to learn. When I reached a place in my life to feel more secure, I started to highlight teammates in-person and online for their families to witness. I would list their accomplishments, their Air Force story, and why I was so proud. More importantly, I was in a place in my life where I was comfortable articulating my appreciation and what I loved about them.

Takeaway — It takes a lot of courage to give someone negative feedback, but it takes a *tremendous* amount of courage to give positive feedback. Are you secure enough to praise others? Leverage your vulnerability and break barriers. Don't just think about something you appreciate about someone, tell them! By helping make those around you better, you are making yourself better as well. I've found that gratitude is simply being grateful for the important things in your life. This might be the missing piece to sharpen your gratitude sword.

Final Thought — The pillars outlined above take a tremendous amount of discipline but are well worth the effort and will to empower you to be fully present for those around you. Neglecting any of those pillars and what they represent can place you in a position that is hard to recover from. Sharpening your sword is an investment for the future. It's the foundation of resiliency.

At its heart, *resiliency* is the ability to withstand and recover from difficult life events.
Rodger Dean Duncan, *Forbes*, 26 Apr. 2022

War Stories

I used to run a lot. (Okay, I jogged regularly. Man, I walked fast and often, okay? Well, maybe it wasn't as often as I used to think.) I was what someone once referred to as a seasonal runner.

My running didn't have much to do with any season, though. I quit whether it was the right season or not. I'd run regularly for a couple of months, and then I'd go TDY for a few weeks, which gave me an excuse to quit for a couple of months—and then I'd do it all over again. It was tough starting over, but it was fairly easy quitting each time. In fact, like any habit, the more I quit, the easier it became. I was soon an exceptional quitter!

I've never been out of shape for long. Whenever I started feeling that way, I'd begin working out. This went on for many years. One day I was teaching Covey's *The 7 Habits of Highly Effective People* to a group of attorneys. As I talked about what I just shared with you, a young captain in the front row got excited and said, "You know, Chief, I have the same problem. If, however, you're willing to be my accountability partner, I'll make a commitment to start a physical fitness program." I'm always up to a challenge. Hey, maybe I could outsmart a lawyer!

We agreed to work out for at least 20 minutes, three times a week, and connect with each other every Monday. We did just that. Every Monday morning, the captain would email me, telling me how it had gone the previous week, and I'd lie and tell her I'd done my part. (Okay, I didn't lie; I really did stick with it.) She had a harder time, due to her workload, I guess. I recall one note she sent after two weeks had elapsed. She'd not met her commitment. Hard charger that she was, however, she said she was on her way to the gym to work out for three hours to make up for it. I immediately called and convinced her she couldn't just make it up like that. The idea was to do it regularly. Amazingly, she agreed. She was eventually reassigned, so I lost touch with her. But I stuck with my program. I'd started riding a bike that goes nowhere fast. By the time we lost our connection, I had developed a habit of working out regularly on that bike. I do it to this day. Or every other day.

Do you know when men (generally speaking) start a no-kidding regular exercise program? After their first heart attack! If you've seen the movie *Something's Gotta Give*, you'll recall that Jack Nicholson's whole perspective on life changes when he has a heart attack. Why wait? Maybe we should attack the heart in a different way—by helping it instead of hurting it. Here's a War Story within a War Story....

My favorite sport is basketball. I have the best seats in the house at Clune Arena at the Air Force Academy, where my Fighting Falcons play. I have my own bodyguard there. I'll call him George. We've had those seats since 2004, so George has become a good friend. Hey! He protects Deb and me! He's a good friend.

A few years ago, we missed George for several games. I, eventually, asked his sub about him. She told us that George had had a heart attack but was okay. He'd been recuperating and would soon be back to work.

When George DID return, Deb and I were elated to see his smiling face again. We hugged and greeted each other as only family does. I asked him how he was doing and what had happened. He reported that all was well and that, in fact, he was now on a consistent diet and exercise program.

NOW on a consistent diet and exercise program? "George," I admonished him, "shouldn't you have been doing that all along? Maybe you wouldn't have had a heart attack." He, thoughtfully, agreed. Can't change the past, but he promised he'd continue on the right path now.

Let me tell you the truth. When I work out, when I ensure that I keep learning, when I'm grateful for myself and the people with whom I share my life, and when I appreciate Earth's unfathomable beauty, I'm invincible! I'm empowered with the strength, wisdom, courage, and grace to accomplish all that I set my heart and mind to. My sword is sharpened, and I know I'll win the battle! I'm ALL IN! BRING IT ON!

Starting Points

• **How can I improve or maintain my physical health?** You can get expert advice on this. I remember calling my protégé, Gary, a retired Chief, to check on him. I ask him what he's doing for a living now. He says, "I yell at people." "WHAT?" I reply, "You YELL at people?" "That's what I do," he tells me. "What do you mean, you yell at people?" I ask a bit bumfoozled. "You see, Bob, I'm a personal coach. I yell at people to do what they know they should do anyway. And they pay me for it. A lot!" He made so much money yelling people that he moved to Florida and bought himself a condo on the beach...cash! You may not be able to afford a personal coach who yells at you, but you know what it takes to get in shape and stay that way. The key is discipline. That's a skill. Develop it now.

• **How can I improve or maintain my mental health?** The best way is to read all of my books! I have a dozen on amazon.com as I write. Okay, read other authors' thoughts, too. I think the discipline in this area of your life is to be disciplined. Set aside twenty minutes a day to read and stick to it. Read material that interests you, otherwise you won't stick with it. And like everything else in your life, the more you do it, the more you'll do it. Study, not just stuff you HAVE to know, but stuff you'd like to learn about. Listen to those Old Farts. They have a lot of experience that you can learn from. As Eleanor Roosevelt once said, "Learn from the mistakes of others. You can't live long enough to make them all yourself."

• **How can I improve or maintain my emotional health?** In the past few years, the science of Emotional Intelligence has grown tremendously. There are plenty of books, videos, workshops, and information on the subject. Google it. As with any learning method, make time to study the emotions that you tend to express and how to manage them. That's first. Then, as you learn to control yourself better, you may be able to help others do so, as well. Controlling is not stifling. But as you learn not only about the emotions you tend to express, but the affect they have on your relationships, you can learn to express your emotions so that they create positive energy.

• **How can I improve or maintain my spiritual health?** Listen. Pay attention. Not so much to others, but to Nature. Nature is a great teacher and it's constantly teaching, awaiting the student. Take a walk outdoors and listen as closely as you can to all the sounds that surround you and contemplate where they came from. I rediscovered my Spiritual Path in Zion, Utah. A very good friend took me there. I was blown away with what I saw and what I contemplated in that place and since. It led me to teach my daughters to contemplate the power of nature. I'd take them to man-made places and ask them, "Who made that?" The answer I taught them to reply with was, "Man." Then I'd take them to places like Sequoia National Park, Yosemite, and Mammoth Lake and ask them the same question. And they learned that the answer was, "God."

Words of Wisdom

Patience is also a form of action.
Auguste Rodin

I have resolved that from this day on, I will do all the business
I can honestly, have all the fun I can reasonably, do all the good
I can willingly, and save my digestion by thinking pleasantly.
Robert Louis Stevenson

*The final forming of a person's character
lies in their own
hands.*
Anne Frank

*Fifty years ago, people finished a day's
work and needed rest. Today they
need exercise.*
Unknown

*A man too busy to take care of his health is
like a mechanic too busy to take
care of his tools.*
Spanish proverb

*The cyclone derives its powers from a calm
center. So does a person.*
Norman Vincent Peale

*Have you ever been too busy driving to stop
to get gas?*
Stephen Covey

Rule your mind, or it will rule you.
Horace

HABIT 6

BE GOOD!
KNOW YOUR STUFF

Remember when Mom used to call out to you, "Be good!" as your little legs carried you out the door as fast as they could? Slamming the door behind you, you didn't pay much attention to her. Habit 6 resembles that scenario to some extent. It doesn't come from Mom, though. This habit has three basic parts. Let me begin by asking you a simple question: What do Michael Jordan, Mia Hamm, Tiger Woods, Brett Favre, Mary Lou Retton, Larry Bird, Magic Johnson, and Kareem Abdul-Jabbar all have in common? Yes, they are—or were—great athletes. What else? Look at the subtitle of this habit. They know their stuff! You may say that some of the "older" folks I listed *knew* their stuff, but that's only half true. The other half is that once you know your stuff, you'll always know it—you just use it in a different way. Older, indeed!

The athletes I listed were great because they were competent. That's the first part of being good. I realize I already mentioned competence previously, but it was in a different context. ***To be good, you have to be competent!*** Oh sure, professional athletes have God-given abilities we call talent, just as you do, but they have to hone that talent. You do that by using it and working on it. Great athletes work hard to develop and maintain their competence. Now that you've graduated to the real world and wear "butter" on your shoulders, do you think that you won't have to work? Sorry to be the bearer of bad news. The truth is that you'll have to work even harder to hone your technical skills because now they matter. In school, what's the worst thing that could have happened if you had failed a test? As an officer/leader, the worst that can happen is that someone might die from your mistake. It might be you. So don't do that!

There is no substitute for competence. I'll share a war story with you that makes that point in a minute or two. A concept that has become very popular during the past few years is readiness—being prepared to go to war. And, let me tell you, if you haven't figured it out by now, the most difficult war to fight is the internal one. The one that wages within yourself. Daily. As a leader, you have to be ready to fight against all the things you'd like to do so that you can do what you must do. And the more prepared you are for those battles, the better you'll succeed. And help others succeed, too. How do you prepare? All of these habits you've been reading about enable your readiness. Specifically, when we talk about being good, readiness begins with being competent to do your job—to know your stuff.

I would be foolish to try to tell you how to hone your particular technical skills. You can do that by getting more training, reading, talking with experts, finding a mentor, or going back to school. (If you're an Air Force Academy grad, you don't even want to hear that one, do you?) Let me tell you that, as a new lieutenant, you will not have time to do those things—I guarantee you. You'll have to *make* time! The world and your work are changing at mega speed. If you don't keep up, you'll be left behind. Do your part to maintain your competence and develop a relationship with your followers that will empower them to keep you informed on new developments they learn about. Remember the web that Chief Seattle talked about? Help your followers help you. *Be competent!*

Being good has to do with character. I can help you with that. (I've been called a character!) I don't know what you do, but I do know who you are. You are a lieutenant. An officer. A warrior. A leader. Your ability to lead is as important as developing and maintaining your technical expertise. When you leave your unit, your followers won't remember what you did. They *will* remember who you were, based on how you treated them. As Maya Angelou said, ""People will forget what you said. People will forget what you did. But people will never forget how you made them feel."

"Being" has to do with character. Chances are you've had some training in character development. If not in school, I'll assume you got some at home from the people who raised you. Since I don't know what you were taught, let me give you the truth. Basically, character means living up to your core values and your corps' values. (Let me commend you to read my book, *Beyond the Little Blue Book*, available at amazon.com, for more insight into that.) Whether it's the Air Force's *Integrity, Service, and Excellence*; the Navy and Marines' *Honor, Courage, and Commitment*; or the Army's *Loyalty, Duty, Respect, Selfless Service, Honor, Integrity, and Personal Courage*, you must *embody* those values. They must come from your personal core—your heart—if you're going to be an Exceptionally Powerful Lieutenant.

Although you're expected to live up to all of the core values of your particular service, the most important is *integrity*. I believe that all the others grow from that one, just as your followers will grow from their experience with you. Integrity is often defined in terms of honesty. That's certainly part of it, but the critical element has to do with wholeness. When a vessel—whether a ship, a tank, or a plane—maintains its integrity, it's whole, and it can accomplish what it was made for. God knows, you don't want to be inside one of those vehicles when it loses its integrity! The vehicles you drive are important, but the vessels you lead are critical. Those vessels are you and your people.

Here's my first introduction to the concept of integrity...way before it became one of the Air Force's Core Values. This happened in the early 1970s.

Chief Master Sergeant (deceased) Jessie Brown was a great musician, leader, and mentor. I don't remember first meeting him, but I remember his lesson on integrity as if it happened yesterday.

I served with Chief Brown in the Philippines, as a member of the Air Force Band of the Pacific, around 1978. We often traveled by C-130 throughout the Pacific Theater. I remember the first time I traveled with the Band. The young Airmen loaded the aircraft before the more senior folks got on the plane. I was a young Airmen then. We loaded the plane, then got off of it through the back to get our personal gear. As we walked toward the front to enter the plane for the trip, I noticed that everyone was hanging around outside, waiting. Not sure what everyone was waiting for, I asked one of the veteran NCOs why everyone was waiting to board. "Chief Brown," was the answer I received. Curious as to why we would all wait for the Chief, I asked a teammate why we would do that. I was told that Chief Brown was afraid of flying. He believed that if the plane were to crash, the tail would be the safest place to be, so he would sit in the very last seat as far back of the plane as possible. Everyone, including the officers, would wait for Chief Brown to board first so that he could take the last seat. I was impressed that everyone respected him so much. On this particular day, he had to take care of some business at the squadron, hence, his tardiness. Once he arrived, and he boarded, everyone followed suit.

For some reason, unbeknownst to me, Chief Brown took me under his wing and mentored me without my asking. I didn't appreciate it as I should have until many years later.

We were on a flight across the Pacific Ocean on an occasion when he chose to teach me the concept of integrity. Remember that this was 1978, decades before the three Core Values were introduced to the Air Force. The air crew used headphones to communicate with the cockpit throughout the flight. Those headphones were connected to the cockpit via a very long cable. The Crew Chiefs would often let us listen in to what the pilots up front were saying. It was a very cool experience. On this occasion, after I'd been on the headphones, Chief Brown came up to me and said to me, "Bob, if you're ever on those things and you hear the pilots say that the plane is about to lose its integrity, grab a parachute...there are only eight!" I kinda got his gist, but not really. He went on to tell me that integrity has to do with wholeness. If the airplane is losing its integrity, it's falling apart. Best to grab a parachute ASAP!

What drives you? If it isn't the desire to do what's right, then reconsider your commitment to being an officer. Doing what's right is basically what integrity is about. To your enlisted followers, being good means doing what's right in all you do, all the time. Okay, I'm giving you the Air Force perspective, but I think it encapsulates all of the services' core, and corps, values. Doing what's right isn't always easy. Michael Josephson told me that character is "doing what's right even when it costs more than you're willing to pay." There's usually a price to pay for doing what's right, but it's worth it in the end. Remember when we talked about discipline as discipleship? Do you think you'll raise disciples if you don't do what's right? I doubt it.

Integrity is clearly a vital component of character development. We often define integrity as doing what's right *when no one is watching*. But what's right? It's not lying, not stealing, not cheating, and not letting anyone else do it either. Most of our military academies include those restrictions in their honor codes. Isn't character about honor?

Maintaining our honor isn't always easy. I think that the most difficult part is "not letting anyone else do it," don't you? That takes what I call moral courage. Turning in a buddy who has violated the code or done something illegal is probably the toughest test you'll have to pass. As a first sergeant, I had to reprimand my very best friend. I would have taken a bullet for him any day (and still would), but as the leader I had to do what was right—and it wasn't easy. Luckily, he had plenty of integrity, which made it easier. "Bob," he said, "the troops are watching us. We need to do what we should. We can't breach our integrity, so let's get this done." He was much stronger than I. We're still brothers. *Having and exhibiting moral courage is key to being a good officer and critical to being a good leader.*

Being good requires having confidence. (Yeah, different context than the last time I mentioned it. I'm preparing you for the test.) In truth, confidence will almost pour out of you if you're competent, if you're a person of good character, and if you exhibit moral courage. You may remember a commercial some years back (if not,

Google it)—I believe it was an ad for a deodorant that warned you to "never let them see you sweat." In a way, that commercial made a great deal of sense. The product would help the user give an impression of confidence. Well, I don't know of any product that will give you confidence, but I do know that if you're competent and have good character, you'll be able to make tough decisions with the confidence that you've done what's right.

Know your stuff! Whether it's in what you do or who you are, make sure you know what you're doing to the best of your ability. You need to know your stuff, but not to impress your followers (they will know much more than you, even if they won't let on at first). You need to know your stuff as a leader and as an expert in your field because everything you do affects someone's life. If you don't know all you should, make sure to invest time in improving your skills. You are not at the top of the heap. On the contrary, you're at rock bottom again. Hey, don't worry. This won't be the first time. Life is a cycle. Once you make it to the top, you have to start at the bottom again. Be good at what you do and who you are.

War Stories

I'm at "that age." There is no definitive number for "that age." I've asked. No one knows. But I'm at "that age." "Chief," my provider tells me, "You're at that age so you need to have this procedure done." Sure. So, I had to go into the hospital for that "procedure." Have you ever noticed how medical folks use fancy terms that don't mean what they mean? "Chief, we need to send you down the hall for some labs." "Labs?" I love dogs. I can envision myself in a room fool of lab puppies playing with them. Maybe it's for my mental and spiritual health. I can do that. Well, when I get to the Lab Room there are no puppies! They're going to suck the blood out of my veins! Tell me the truth! Okay, maybe not. I might not go if I know the truth. A "procure" is when they invade your body with an instrument much larger than the orifice they plan to stick it in? Have you ever had your ears cleaned? See what I mean? Okay, so I'm going in for this procedure that a "man at that age" should have. Or so I'm told.

Now, I really needed this procedure, so it takes only four months to get an appointment to talk to a doc about it. (Yeah, I go to a military hospital.) It would eventually turn into five months before I had fun. Therefore, the man my age was now a lot older. Probably at "that age." The doc's a good guy. He'd served 30 years before he started getting paid for what he did in the service for free. In other words, he's a retired GI like me. I'm very interested, to say the least, in what the procedure entails. I'm sure my blood pressure has increased as he explains what will happen. After he's done, he asks me if I have any questions.

"Doc, how many times have you done this?" I ask him. "About 24,000," he answers, to my relief. "Has anyone ever died while you were doing it?" I ask. "Not yet," he says proudly. YET?" Doctors shouldn't try to be comedians. Imagine if he'd answered my question about how many times he'd done this procedure with "Oh, this is my first time, but don't worry. I graduated at the top of my class, and I've read all the manuals!" No way! You ain't practicing on me, man! I suppose you have to start somewhere. But not here! Although I can't verify this, I've often heard that medics who work in immunizations practice on themselves. Some things you can't practice on yourself. You'll find that out in due time, when you reach "that age" and need that procedure.

You have to know your stuff. Would you go to a doctor who hasn't been trained, one who doesn't know what he's doing? Of course not! Why not? Because he might kill you! It's the same in your business. You have to know your technical stuff so that you don't get someone killed, and you have to know your character stuff so that you don't destroy someone's life, including your own. And you have to be confident that you've done what you should. If you noticed, as I described the doc who worked on me, I first mentioned that he was a good guy. I first tried to check out his character, but I also made sure he was technically competent. Based on how he answered my questions with great confidence, I felt that I could put my life in his hands. I survived the procedure, although he admonished me to see him again in five years. Well, if I ain't dead by then, maybe I'll go have more fun again. Be good! Lives depend on it!

Here's a story about competence shared with me by one of my protégés, Chief Master Sergeant Joe Bogdan.

In his book *The Speed of Trust*, Stephen M.R. Covey explains that trust is comprised of both competence and character. Many feel that trust is just a character issue. However, you can be a person of integrity and kindness, but if you aren't competent at what you do, you won't inspire trust and confidence within your teams.

On my last deployment, I was the Senior Enlisted Leader over a diverse unit of 26 different Air Force Specialties comprised of warriors from support, operations, and maintenance backgrounds. We were in Al Asad Air Base, Iraq and just 10 months prior, the base was hit with a barrage of Theater Ballistic Missiles (TBM), the largest ballistic missile attack ever against Americans. On my team, I had a Company Grade Officer who was the Flight Commander over 50 Security Forces personnel responsible for defending the Airfield. We were battling COVID as well as Shia Militia Groups and remnants of ISIS, while also in the midst of a slow troop withdrawal in the country. It was a tense time, and the entire unit was stressed.

The Security Forces team had a lot of personnel issues and the Flight Commander struggled with upholding standards. On many occasions, he relayed bad information to us as a Command Team, and we began to question his competence. We also started to notice more and more breaks in discipline among the personnel in his flight, and it became apparent that his personnel did not have confidence in his professional, technical, and leadership competence. The young Master Sergeants that were assigned to his team did their best, but they also had their own shortcomings. Then, one night we had an inbound TBM warning, and his team reported that he immediately sprinted for cover alone and with no regard for the safety and security of those whom he was charged to lead.

In the following months, it was clear that his team did not trust him, and we lost faith in his leadership as well. His lack of competence killed his credibility and continued to erode at the good order and discipline of his flight. If it wasn't for a few strong Technical Sergeants, we would have had more problems within that flight and unfortunately, with such a limited bench during our downsizing, we did not have an option to fire someone even in a contested environment.

A leader that isn't competent at any level within an organization impacts the entire team. We as a Command Team had to continuously monitor and re-vector this individual's decision-making and found ourselves constantly dealing with personnel and disciplinary issues within that flight. This took our attention away from the other flights and wasn't fair to the rest of the unit. Additionally, the team's performance reflected their poor leadership, and this ultimately impacted the security of our airfield and our overall mission.

We should all seek to be an expert in what we do...it matters! Start off a cook but strive to be a chef! Don't just try and speed to a promotion, but, instead, master the skills that the promotion will require so that when you get promoted, you deserve it. Just like you wouldn't trust a doctor who skipped steps to get there, you shouldn't expect that someone would trust you as a leader if you did the same. Master each step and KNOW YOUR STUFF!

Starting Points

• **Do I really know what I'm doing?** Chances are that the answer, honestly, is NO. But that's almost okay. One of the keys to being a good leader is self-assessment. You've probably heard that before. I'm not sure we do that much or well. The next three questions may help you with that, but it's important to take, or make, time, regularly, to ask that question and after you've assessed your strengths and weaknesses, devise a plan for improving yourself. The answer to this question should be preceded by asking, "Do I know what I think I should be doing?" Both will be follower-centered in that your goal as a leader should be to help your followers succeed. Help them empower themselves to be their very best. That's what an Exceptionally Powerful Lieutenant does.

• **Do I need more training?** Surely, you do. You may become an expert at what you do, but leading is more about who you are and that will be a continuous process of growth, if you do it right. Get as much training as you need to be good at your job. Keep in mind that your followers are watching you. If you continuously strive to improve your skills, they'll do so, too. A learning organization is a growing organization. As a leader, you'll never know it all. NEVER! As soon as you think you have your followers figured out, they'll change, or grow, and the process will begin, or continue, again. That's okay. That's what's fun about leading. There's never a dull moment. Keep growing. Keep learning. Keep practicing. Keep caring.

• **Do I embody my corps' values?** This question may be tricky to answer. You may THINK the answer is yes, but is it, really? I'll share more on this in Habit 8, Hang on Tight!, but what's going to make you exceptional is not only your assessment but that of your followers. As I said above, self-assessment if critical to being an effective leader, but there's huge value in knowing what your followers think about you and your leading abilities. Find an honest, trustworthy, Senior NCO who will tell you the truth about how your followers see you. Find a peer who you can trust to be honest with you about how your fellow officers/leaders see you. Be careful not to beat yourself up when trying to improve. No one is perfect, but the more you strive to live up to what you all value in the profession, the more your followers will stand behind you and next to you.

• **Do I continuously do what's right?** You will be tested. Continuously. You're the leader. You're expected to do what's right. It may be difficult at times, but you'll have to do what's right, not what's popular or easy. Welcome to leadership! Answering the previous questions will set you on the right path, assuming that your answers are in the affirmative, and propel you toward getting better no matter how good you already are. Strive to be your best at everything you do. Notice that I didn't say THE best. As Theodore Roosevelt once said, "Do what you can, with what you've got, where you are."

Words of Wisdom

The final test of a leader is that he leaves behind him in other men the conviction and will to carry on.
Walter Lippmann

Say what you mean, mean what you say, but don't say it mean.
Anonymous

*Never let a problem to be solved become
more important than a person to be loved.*
Barbara Johnson

*Nearly all men can stand adversity, but if you
want to test a man's character,
give him power.*
Abraham Lincoln

*Doing the right thing for the right reason in
the right way is the key to quality of life.*
Stephen Covey

*Quality is never an accident; it is always the
result of high intention, sincere effort,
intelligent direction, and skillful execution; it
represents the wise choice of many
alternatives.*
Willa A. Foster

*You cannot dream yourself into a character;
you must hammer and forge yourself one.*
James A. Froude

Talent is a gift, but character is a choice.
John C. Maxwell

HABIT 7

BUILD TRUST!
BE TRUSTWORTHY

At the beginning of Habit 6, I asked you to relive your days at home when Mom told you to be good. Now, I don't have the female perspective, never having been one, so forgive me if I sound sexist, but here's something all men go through. Recall, if you will, the first time Dad let you have the keys to the car so that you could go out on a date. Did he give you the "I brought you into this world, and I'll take you *out* and make a better one—and it'll be more fun this time!" talk? In case he didn't, let me play that role with you now.

Dad: "I'm trusting you with this vehicle that I've worked hard to pay for. (All dads are trained to say things in a way that will make their children feel guilty enough that they'll do the right thing.) I've sweated countless hours so that the family would have a decent car to travel in. Now, I'm going to trust you to take care of it. I know, however, that sometimes things happen. (Dads are also trained to practice tough love. The idea is to incorporate guilt with responsibility, hoping to produce integrity.) If something should happen to the car—

for instance, if someone runs into you (notice he didn't say if you run into someone; that's different, and you will pay for it!)—I expect you to tell me about it. (Dad is saying, Keep me informed. (You'll see that again soon.) I'd rather you tell me about it than hear of it from someone else. Don't ever lie to me. (Now get ready for the coup de grâce.) I can deal with an accident happening; it happens to all of us. ("Yeah, right," you're thinking. "He's going to understand!" A good dad will.) What I won't put up with is your lying to me about it. Understand? (Here's where he adds the "I brought you into this world, I'll take you out" for emphasis.)

Of course, you understand, and you tell him so (although you hadn't yet read this book back then, so you didn't know how to listen—consequently, you shut him off eight seconds after he started).That conversation has to do with building trust. ***Trust is the most important thing you'll have to develop among your followers if you're going to be an Exceptionally Powerful Lieutenant.*** Trust binds people together so that they can accomplish whatever mission lies ahead.

Here's a critical question to ask on this topic. Who do we trust? Wait a minute! Before we get to that, let me ask you to define *trust*. Put the book down a minute and think a little bit about how you define trust. And since you're thinking, consider who you trust....

Thanks for coming back. What comes to mind when I ask you to define the word, *trust*? I'd bet my huge retirement paychecks that you said something to the effect of "trust is when people are honest" or "it's when people do what they say they will do." Am I close?

I always begin my Building Trust workshops by asking my audiences to define or describe the term, *trust*. Typically, I'll get those answers I just shared with you. If you'll notice, they're externally centered..."trust is when someone else does something." Sure. But to be an Exceptionally Powerful Lieutenant you'll have to develop a different perspective. Before I share that perspective with you, let's go back to my original question, "Who do we trust?"

We trust people who are competent, who are confident, who keep us informed, who listen to us, who are considerate (who care), who make themselves available, who are consistent, and who are principle centered. Let me tell you that trust is built from a combination of these eight traits. You can't build trust by doing only one of them. The more of them you practice, the more trust you'll build. Let me address each one separately.

We trust people who are competent. Habit 6, Know Your Stuff, was about the importance of being technically and morally competent. I think you understand the importance of being technically competent. If you don't know what you're doing, then, maybe you should either not do it, or learn how to do it correctly. You can figure that out, I'm sure…or ask a Senior NCO for guidance. But knowing isn't enough. One of my most favorite truisms is, "To know but not to do, is not to know." I'm not sure who said that. I wish I had. The real power in your leading ability will be in passing on what you know. That's real HEIRPOWER! And not just telling others what you know but showing them. In the Foreword of my book, *What I Learned From Dad Made Me a Better Man!*, I mention that "We cannot teach character. We cannot teach leadership. The best we can do is show it. People will then follow." That's the truth! Your followers may not always do what you say, but they'll do what you do. Express your competence in your behavior more than your words.

Personal Competence requires self-knowledge. I mentioned it briefly in Habit 6. I even mentioned it in the Starting Points section. (Yeah, this is a review. And there WILL be a test!) Know yourself. You're made up four basic domains. They're your values, your purpose, your vision, and your influence. Take some time to ask yourself the following questions and answer them as honestly as you can. If your answers make you tear up, that means you got into yourself very deeply. Do it periodically. I do it between Christmas and New Year's, every year.

What's truly most important to me? What are you willing to die for? God forbid you ever have to do that, but if you had to, what would that be? I'm referring to your values. Most of us think we know what's important to us, what our values are, but until we've written them down and made a conscious decision that they truly are your values, what's most important to us, we really don't know.

Why do I do what I do? If it's for the money, find something else to do. Money won't buy you happiness. It won't! As a leader, if you do it to help others empower themselves to be their best, your followers will call you Leader. Once you've discovered your purpose, you'll live up to it in spite of the challenges. Maybe even appreciate the challenges.

How do I see myself and my followers? If you see yourself as a capable caring leader, you're more likely to be one because you have a vision of what that looks like. Make time to envision your followers at their very best and help them become that vision, or better yet, their vision of their very best selves. When you envision your followers as great, you'll tend to treat them as such. And as you value them, they'll value you.

How am I influenced and how do I influence others? Yeah, I know I'm asking you to think deeply, but it's important to know these things. Once you've assessed how you're influenced, you can choose to continue to be, or not. That's real power. The power to choose! There are things and people you'd probably prefer to be influenced by more than others. And never forget that your behavior, good or bad, WILL influence your followers. I know I mentioned it in Habit 6, but it bears repeating. Maya Angelou said that "People will forget what you said. People will forget what you did. But people will never forget how you made them feel." There's a really good book you might want to read that focuses on these concepts. It's titled, *The Power of SUPERvision!*. It's on amazon.com. I wrote it. It really is good! Trust me!

I suggest you go through this self-assessment at least once a year, as I do. The answers may change or evolve, and that's okay. But as the great philosopher, Yogi Berra, said,

"If you don't know where you're going, you'll end up someplace else."

Competence, especially as a leader, requires that you know your followers. Recall my War Story about being present for your followers in the previous habit. (Yeah, more review.) Being present is the first step toward knowing your followers. You can't learn about them unless you're with them. Yeah, yeah, you're invoking technology as a way of learning about them. Technology is always limited and limiting. Your followers aren't. They have a lot more to give than you know and that they'll tell you. Unless they trust you. Being present is a gift! Yeah, I said it! High tech will never trump high touch!

For several years, I worked with one of the most humble and caring people I've ever known. Often, he'd come in to work later than everyone else. Doing the Guy-thing, I'd give him a hard time about coming in late. If you're a guy, you know what I mean. Nothing malicious, just being funny…kinda. One day, as I was admonishing him for his tardiness he says, "Yeah, Chief. Last night was a particularly hard night with our autistic son." WHAT?

"You have an autistic son?" I ask him.

"Yeah. He's six and sometimes doesn't sleep all night. My wife and I take turns watching him to make sure he doesn't do anything to hurt himself. That's, often, why I come into the office later than everyone else."

AAAWWW, CRAP! You can imagine how I felt finding out this knew information. As I said, he's one of the kindest people I know and I'd been giving him a hard time, supposedly in a jovial manner, about a HUGE challenge in his, and his family's, life. I, of course, apologized for being a jerk. He, of course, accepted my apology in a very gracious way.

Chief of Staff of the Air Force, General Mark Welsh, made a speech at the Air Force Academy that went viral. It was about the importance of knowing your followers' stories. You can find and download the entire video clip of it at the following link, https://www.youtube.com/watch?v=jxFELfMZlmg, in case you haven't seen it. Near the end of the clip, the general mentions that every Airman has a story. That's true, but consider this… Every Airman IS a story. Your mission, if you choose to accept it, if you aspire to be an Exceptionally Powerful Lieutenant, is to know that story. As General Welch so powerfully declares, "If you don't know the story, you can't lead the Airman."

Often, leaders don't know what they don't know. Find out! Know your followers!

Competence creates confidence. Let me make a point on how confidence builds trust. Imagine if that doc I told you about in my war story came into the operating room scared and wondering if he could do what was expected of him that day. How long do you think I would have stayed on that operating table? I'd have been out of there in a New York minute! It's the same with leading your followers. Will they follow you if you're incompetent? Nope! Will they follow you if you're indecisive or show fear? Of course not! A scene in the movie *U-571*, which I'll refer to again in Habit 8, illustrates what I'm talking about. The scene depicts a conversation between a crusty Navy Chief and the Skipper. The point in that scene is that ***we trust people who are competent, and we trust people who are confident.***

You have to be confident in who you are if you're going to lead others. People will not follow a leader who is not confident. Now, there's a difference between confident and cocky. Don't be cocky. No one will follow you then. Confidence requires humility. We often confuse humility with ego. There's a huge difference between a strong ego and a big ego. People who have a big ego go around telling others how great they are. People who have a strong ego don't have to tell anyone anything, they just show how good they are and share their talents with those they lead.

Here's the bottom line…if you see yourself as weak, as incompetent, as a dirtbag, you'll fulfill that prophesy. You'll act like that and soon become it. If you see yourself as strong, humble, competent, as a leader who helps others empower themselves to fulfill their purpose, you'll become that the more you practice it! It's your choice.

Back in the day, we were taught to work on our weaknesses until they became strengths. I've been there and tried that. It doesn't work! If you're naturally weak at some things, all the work in the world won't make them strengths. New thinking suggests that you should work on your weaknesses so that they're not deficits but invest yourself in developing your strengths. Do what you're best at and hire someone to do what you're weak at. You may even help others in your strength areas that are weaknesses to them. That's win-win thinking.

We see others more clearly than we see ourselves. Why? Because we see ourselves by our intentions. We see others by their actions. We may mean to do something, but not do it. Others did do it. Leading effectively begins with YOU! How well do you know yourself? How do you see yourself? See yourself as a Shining Star! See yourself as competent and confident. And then you can see your followers as such, too.

Know this: Everything we do is purposeful! EVERYTHING! What's your purpose for breathing?

210

So that you don't die! What's your purpose for sleeping? So that you can maintain your mental health! What's your purpose for eating? You catch my drift....

Since my friend, Simon Sinek, wrote his book, *Start With Why*, which, by the way, I commend to your reading, many, many people have referred to the notion that starting with why, which, Us Old School Cats, used to call "purpose," is important to success in any endeavor that you undertake. That's partially right. Here's what's critical, though. Is it the RIGHT why?

Back to my question about eating. What's your purpose for eating? To sustain your physical body is the MAIN purpose. Is that the right purpose? It is! But what if you eat because you're nervous or stressed? Is that the right purpose? I'm not judging here. I'm encouraging you to think deeper than just your purpose for eating, or for doing anything, for that matter. Is it the RIGHT one? Think about those people who enter the annual hot dog eating contest. Is their purpose different than that of why you and I eat hot dogs? Probably so....

Everything we do is purposeful and values-driven. Finding the right why starts with discovering, often, RE-discovering what's important, MOST important, to us. How do you know what's most important to you? I've read that reference to your calendar and your check book will tell you. We used to pay bills with checks. Google it. Where do you invest your time and your money? What do you buy? What do you pay for? What do you do? What's most important! You will always do what's most important to you at the moment. What's most important to you at this moment? Reading this! How do I know? Because you're doing it!

Here's an easy test of what's most important to you.... How do you start your day? What's your first thought when you wake up? You can make that intentional, you know. When I wake up at 0300, without an alarm clock by the way, my first thought is that of gratitude. I thank the Creator for my family, my friends, my blessings, and myself. Then I go pee! Hey! It's a natural thing four times a night at my age. Then I go about doing what I should to take care of those blessings I'm grateful for.

Before you start with why, make sure it's the RIGHT why. Think about what and who is most important to you and how you can affect that and them in a positive way TODAY! You only have today. Find your RIGHT purpose, your RIGHT why!

Your purpose, the RIGHT purpose, will propel you to have and exhibit the confidence you'll need as a leader. Your followers will know why you do what you do and that you do it for a higher purpose than just making money or being cool. When that happens, guess what they'll do? They'll follow you. Competence creates confidence.

Cadet Bratka was one of the largest people I've ever known. I don't know exactly how tall he was. I'm going to guess 6 foot 12. And he weighed a ton. He was (probably still is) solid muscle. When I patted him on the back, my hand hurt.

So, I'm on the Terrazzo at the Air Force Academy, heading toward the Ramp. For those who don't know, the ramp leads to freedom from the Terrazzo, which is a flat square a few acres in area. (That's as close as I get. I'm not a mathematician.) Anyways, as I'm heading to my destination, I begin to feel the ground shaking. But it's in rhythm. And this is Colorado, so it can't be an earthquake. As I get closer to the Ramp the shaking gets stronger. I look down the Ramp and notice the head of a cadet climbing up the Ramp toward the Terrazzo and me. The head becomes shoulders, then torso, then a full-sized person, otherwise known as Bratka. He's trotting toward me. That's the shaking I feel. It reminded me of when Mongo rides into town in the movie, *Blazing Saddles*. As Bratka gets closer, I notice he's calling out to me. "Chief! Chief!" he's calling. I acknowledge him and head toward him... very slowly.

He comes upon me and picks me up with a huge bear hug! "Chief! Chief!" he reminds me of who I am. "I did it, Chief!" he announces.

"What did you do, Bratka?" I ask, barely breathing, "And put me down, Man!" I admonish him.

"I did it Chief! Remember that little card you gave us at VECTOR!? (The workshop I used to do for all of the freshmen cadets.) I DID it! The card asked how we saw ourselves in the future. I wrote down 'Wing Commander' on mine. The Comm (Commandant of Cadets) just told me I'm gonna be the Wing Commander next semester! I DID it, Chief! THANK YOU!" Needless to say, he was elated with the news. He was proud of himself. Almost as proud as I was of him.

For the past few decades, I've admonished Shining Stars that "If you can see it, you can be it!" I even wrote a book based on that thought. Develop that self-confidence so that you can lead effectively.

Recall that in my opening "Dad talk," I promised that you'd see "keep me informed" again. As a man of integrity, I have to live up to that promise, so here it is. The whole point of that conversation was that **we trust people who keep us informed regardless of the news, whether it's bad or good.** Truthfully, as a leader, you should have more trust (if levels of trust exist) in those who give you bad news. Whom would you rather hear bad news from—one of your followers or the general? I rest my case. That's the point your dad was trying to make.

You've heard the adage that "information is power." That's bunk! Information can't think! People think! Sometimes.... Information doesn't care about power, it's just information. Shared information is powerful! The dictionary defines power as "the capacity to act effectively." Does information itself have capacity? No! People do! Power lies in people and how they perform. Power is increased when information is shared. Once again, how do you, as a leader, make decisions? It's based on the information you have at hand. If you don't have all of the valid information you need, you'll end up making bad decisions. We trust the people who keep us informed, especially those who provide us the bad news. Every colonel and general officer I ever worked with admonished me to keep them informed, especially of the bad news. Every one of them chartered me to keep them out of trouble by telling them when their decision might be a bad one based on the information I had. They knew I'd have different information than they were given.

I remember General John Jumper, then Commander of the United States Air Forces in Europe, asking the USAFE Band, to which I was assigned, to develop a tattoo for Ramstein Air Base to be performed on the Air Force's birthday, in September. A tattoo is a military ceremony, usually done in the evening, with music and marching. It's a very exciting and interesting ceremony. This was a very high-level event, especially since the Big Boss had personally asked for it and many other Big Bosses were expected to attend. You can imagine the amount of effort it took to coordinate everything.

After months of work, we were notified to ensure we had a rain plan because the weather folks expected rain on the day the tattoo was scheduled for. Duh! It always rains all of September in Germany. Anyway, we thought we had that in the plan. The performance day arrived and, lo and behold, we awoke to constant showers. The weather guys were right for once! Panic, everyone! The sky is falling! Literally! As soon as we got to work, we got the call to get to a meeting to work out what we were going to do. The rain plan called for taking the tattoo indoors. Not a problem, you'd think…if you've never been stationed at Ramstein. Ramstein has to be the busiest base in the world. Okay, you think what you want, but it's the truth. The hangar we'd planned on using in case of rain, back a few months ago when we developed the plan, now had a C-130 parked in it. In pieces. I know, your first thought is to move it out. No can do. Like Humpty Dumpty, all of the general's men couldn't put it together again in time to move it out and prepare the hangar for the tattoo.

Aha! There are other hangars on that base. Let's use another! The others, too, weren't available. After much cussing and discussing, someone suggested we get in touch with the Logistics Group Commander to find us a hangar. Hey, at least we could then blame him for not making it happen and we'd be clear. The LG told us none were available. Someone at the table said something about going by Hangar 3 and seeing it empty. We called the LG back and told him what we'd heard. Nope. It's got airplane parts in it. It was about lunch time now, so we decided to take a break and return in an hour. Being the sly and cunning enlisted guy that I was, I decided to stop by Hangar 3 to check it out and maybe come up with other options. As I walked into the hangar, a Master Sergeant came up to greet me. He was smiling a huge smile because I'd come by. After some chit-chat, I asked him if he was expecting to get some gear in because the hangar looked empty to me. He said, "No, Chief, we just cleared it out." You know what I did. I called my boss so that he could get the credit for fixing the problem. (My boss did get promoted that cycle, by the way.)

We did the gig and it went tremendously well in spite of the weather. The point is that the LG, who was a great leader in his own right, made his decisions based on the information he had. The Master Sergeant hadn't had time to tell anyone his crew had completed its task way ahead of time. He had the information but hadn't shared it yet. Sharing information is powerful! It provides us the capacity to act effectively. I think you'll agree that we trust those who keep us informed.

I walk into my home and see my lovely bride, Deb, smiling. "Hey, Baby," she says. "Hey, Baby," I reply, "How are you?" We go on, asking the usual questions about each other's day, how it went, what we did, etc. Then we get to a critical part of the conversation.

"You know," Deb says, "I was talking with Lori today." Lori was married to one of my colleagues. "Oh?" I asked, "what about?" "Well, we were just chatting when she mentioned something about you guys being gone for Thanksgiving."

I was on active duty when this occurred. I was assigned to the United States Air Forces in Europe Band at the time, stationed in Germany. One of the coolest things we did in that band, and I take at least partial credit for creating the concept, was that we deployed a big show we called Seasons' Greetings to locations where our troops were. We had great success with those shows because we brought in artists from the US and did a big production for our troops and their families, similar to a USO show. I'd had this great idea of playing for the troops (no families there) stationed in Bosnia for Thanksgiving Day! I'm sure you're saying, "What a GREAT idea!" Everyone I'd talked with said the same thing, "What a GREAT idea!" With one exception….

"Lori told you we'd be gone for Thanksgiving?" I asked Deb, trying to sound as if...okay, the OC Factor kicked in. "Oh, Crap!" was all I could think of. In my zeal to get the show together and make it a great event for the troops, I'd forgotten, no kidding, I'd just forgotten, to tell the most important person in my life what I was up to and that it would affect her and the girls since I wouldn't be there for Thanksgiving dinner.

That old saying, "information is power" is stupid! Information is just information. It won't get up and do anything powerful! It'll just sit there, sometimes in your brain, and do absolutely nothing! Imagine being the smartest person in the world and no one knowing it? What good would that be? SHARED information, APPLIED information is power.

Sometimes people who call themselves leaders think that keeping information from followers is power. That's REALLY stupid! When we empower others with the information we have, it makes us ALL more powerful and it builds trust.

If you're looking to be effective, whether it's at work or at home, ask a critical question often, "Who needs to know?" Don't you make decisions based on the information you have at hand? Well, imagine having all the information you need! Where would THAT come from? If you have people keeping you informed, then you're liable to make better decisions.

And the law of reciprocity says that when you inform others, they, in turn, inform you. You get what you give.

If you're wondering what ever happened that Thanksgiving, we did do the tour. In fact, there's another story I'll consider sharing with you about a very cool thing that happened on that tour. But not here. Deb and the girls understood my purpose and they supported me being gone that Thanksgiving. And I'm almost done paying for that BMW I had to buy Deb to make up for it. It's a very nice car, I must say. SHARED information, APPLIED information is power. ***We trust people who keep us informed.***

I already gave you my perspective on listening in Habit 2. Does it make sense that ***we trust people who listen to us***? (Go ahead, "Nod your head," as Sir Paul McCartney would say. Google it….) In Habit 2, I shared some thoughts on the importance of listening empathically and how to do it. But what about why? That's the one question your followers will always ask, "Why, LT?" It's universal. Think about this as you answer that question. (By the way, NEVER answer, "Because I said so!" or "Because I'm the boss!" Those answers NEVER work. Let me correct myself. They MAY, in fact, work…negatively….)

Deb's niece (I don't know what relation that makes her to me), Alexis, posted a very good article on her personal blog the other day, https://www.lexbereal.com/, which, by the way, is pretty cool, titled "Is technology making me lonely?" She's a young person. I'm not! So, I have this Old School perspective, which, by the way, you're developing daily. If you're not yet, you'll soon enough be considered an Old Schooler. Trust me. It will happen before you know it. Soon! We Old Schoolers, constantly bash the Youngsters for being who they are. You may already do it, which is the first sign. I know. I need to GET OVER IT! But maybe this will give you a perspective you hadn't considered.

When I was a Youngster, and I was, just as many of you were, back in the day, we used to go to the library to "study." Sometimes we did. Study. The main reason we went there, okay, second reason, was that the library was a repository of knowledge. There were thousands of resources there that were called books! Most of us, at least in my barrio, couldn't afford books, especially that many, nor about all the subjects the library could. We were hungry for knowledge. We needed it to get the papers done that were required in our classes. Some of us even read some of those books for pleasure or just to gain information on topics of personal interest.

Somewhere along the way, there was a turning point where that knowledge that was in those copius books was supplemented by a new thing called the internet. What a game changer! Now, we could access even MORE knowledge, and much much faster! I could never stand those card catalogs. I never figured out how to use them. It was much easier to ask the librarian, who knew EXACTLY where that particular book was. EVERY TIME!

Anyway, when the internet came into existence EVERYONE wanted to get on it. It, however, was not accessible to most of us. Either it wasn't available, or we couldn't afford it. Our conversations went like this… "Man, I need to get on the internet. How am I going to do that? Where can I get on it for free? THE LIBRARY!" The library was the first place to have the internet for FREE! That's how I remember it, anyway. So, we hung out at the library so that we could connect to the internet without paying for it. We were still in search of knowledge.

Another turning point occurred when Steve Jobs created the iPhone. I recall the Dean of Faculty at the Air Force Academy, Brigadier General Dana Born, challenging me to consider what the library of the future would look like. I took out my iPhone, showed it to her, and told her, "Here it is!" (Interestingly, they renovated the library there. There are no books in it!)

I was in our library just after arriving at the Air Force Academy, talking with some Freshmen Cadets, when the topic of information and knowledge came up. I suggested to them to look into all of the books around them. They all gave me a bumfoozled look, looked around at the shelves, then acknowledged that there actually WERE a bunch of books in their presence. They hadn't noticed them until I pointed them out.

Youngsters now have the entire world's library in their hands. LITERALLY! There's no question we Old Schoolers can devise that they can't answer within seconds. And they'll have tons of answers! Now, the answers may change even as they Google the question, but they have immediate access to the latest facts.

So, here's my point. Youngsters don't need to go to the library to search for information and knowledge in books anymore. They don't have to go there to connect to the internet, they can do it wherever they are, except in my office. That's another story. So why do they go to the library? They still do, you know. At least my former 18-to 22-year-old charges do. They go because they're looking for relationships!

Facebook may be a great source of knowledge, but what they, and we, are looking for is relationships. Why do they call it Facebook FRIENDS or Facebook FAMILY? It's all about having relationships with others!

What Alexis says in her article, in essence, is that technology that SHOULD bring people, Youngsters and Old Schoolers alike, together, often, doesn't. It separates us. You see, before High Tech we had High Touch. We knew how to talk with people back in the day. That's how we communicated and related. Our conversations were real time. They occurred at the moment. We REALated! (I just made that up!) Now, Youngsters don't have the skills, because they don't have the training, to relate to others. The communication they practice, or think they do, isn't real. It's captured on an electronic device. That's convenient. You can read whatever anyone said at your convenience, not when they meant to tell you. What's worse is that they believe it's real. If it's not happening right now, it's not real. I'm referring to communication, of course.

So how do we Old Schoolers fix it. That's what Old Schoolers want to do, you know. Fix THEM! Maybe we should model REAL communication for them. Maybe we should, no kidding, LISTEN to them. Show them what that feels like. My sense is that they'll like it. Maybe they'll even emulate it. Youngsters, by the way, see us Old Schoolers as role models. They won't admit it, but they do.

One of my favorite adages goes something like this: "We get so hung up on giving our kids what we didn't have that we forget to give them what we DID have." We used to relate to, and with, each other in real time. At the moment. We used to share our feelings in our communication, and by that I don't mean coddling. We could share empathy as well as sympathy. Even anger. But it was real in real time. We understood better because we could actually SEE the feelings being expressed by the messenger.

Your followers are people. They're great people with the potential to achieve great things and to be great citizens. If you don't help them learn to communicate as humans, who will? They won't know what they don't know, but you do. Do something about it today. Start with asking them how they're doing, listen for their answer, and act accordingly.

The reason for listening is to develop relationships. Professional relationships. Building trust creates relationships. That's the foundation of trust. And leading is all about creating, developing, and maintaining strong relationships. But you have to be trustWORTHY. In the listening realm, that begins with knowing why, or the purpose, for doing it, and why your followers need it.

In Habit 4, Be Care-full, I shared with you that *your followers will trust you based on how you care.* Once you care...and you show it...they will follow you....

I've known and worked for many "leaders" who SAID they cared but didn't show it. Which means that they really didn't care. Or maybe they did but couldn't show it. But they never asked for guidance as to how to show it. OR maybe they didn't care enough. Now, you're asking what that means, caring enough. If your followers don't believe you care, then you don't. Caring isn't about what or how YOU think as much as it's about what and how your followers think. To show you care, you have to know what that means to them. Then, choose to behave in that way, or not. It's always your choice. I hope you're ready for another story....

Man, I was so proud of myself! It was about noon when I got home and as I walked into the house, I found my lovely-bride-of-16-wonderful-and-fulfilling-years, Deb, and my 10-year-old daughter, Tesa, having a discussion. Okay, they were arguing. Oh, all right, they were fighting! I didn't know, nor did I really care, what they were fussing about. I had a very important and special announcement to make! I quieted them down and proudly announced that I had just been selected to be the Commandant of the March Air Force Base Noncommissioned Officer Academy. That was such an historic event that I left work two hours early! I gave up my daily tee time to go home to share my excitement with my family! I'd worked so hard to reach that pinnacle of success! No other bandsman had ever done that! "Man, I'm good!" I thought. "I've made it to the top!"

As I expected, I could see the pride in Deb's eyes. Tesa, our first-born, looked at me as intently as I'd ever seen her look, and without a moment's hesitation asked, "Fine, Dad, now will you help me pick up poop?" Oh, as you may imagine, my ego immediately deflated to nothing.

You see, although my career was of the highest importance to me, what was much more important to Tesa was what I often considered minutiae. It's the little things that our children and our families find important. Often, our children won't know, nor care, what we do as a profession, but they'll always know whether we were there when they needed us. And, trust me, they will always remember. Tesa still remembers, decades later.

I did go out and help Tesa pick up poop. Afterward we sat down on the stoop, and she said to me, without any prompting, "Dad, I'm proud of you." I never asked her if she was referring to my news or my help. It didn't matter. She was proud of me and that's all that mattered that day.

Pay attention to what your followers need and be there for them. And be there for them even when you're not there. More on that soon. Show them that you care in their language, not yours.

I'm pretty sure I already mentioned the concept of humility to you. It's critical to being an Exceptionally Powerful Lieutenant. Practicing it shows you care. Including your followers in decisions you make, as much as you can, is a form of caring that will take you a long way.

"What do you think?" Asked sincerely, with the right motive behind it, this is one of the most powerful questions a leader asks. I recently encouraged you to think about the importance of being valued and valuable, two things that are critical for your followers to feel. Asking the question, "What do you think?" implies that you value the person's opinion, hence, you value the person.

If you're in a leadership position and don't ask that question, your chances of failing are huge. Who know what's going on in the organization? The leader? Sure...! THE FOLLOWERS! The workers who are on the floor, in the mud, know a lot more than you do! It might behoove you to seek their opinion once in a while, or, better yet, as often as you can.

People support what they create. No involvement, no commitment. If you aspire to be a good, or even great, leader, you have to include your followers in your decisions as much and as often as you can. You're still accountable for the final decision, but when you ask your teammates for their opinion the final decision will be partly theirs. That's the essence of teamwork. Working together toward a common unified goal. What do you think?!

What about availability? Let me go back once more to your parents. I could be getting into a touchy area here. Forgive me if I do. I told you from the beginning that what I share with you is the truth as I know it. One of the greatest blessings I had growing up was that I had two parents. I'm oh so grateful for that! Not only did I have both my parents, but they were always there when I needed them. I can recall the enthusiasm with which Mom and Dad went to all of my concerts when I was in the band in junior high and high school. Have you ever heard a junior high or high school band? Okay, let me not offend anyone more than I have to. I'm being cynical when I refer to my dad's enthusiasm. But he was always there! He was always available, just as Mom was. I could always count on being able to talk to my parents about anything. Often, they had no clue what I was talking about, but they would sit and listen. I hope you grew up in that type of environment. If you did, you know what I'm saying when I tell you **we trust people who are available.** Imagine having an open-door policy, but you're never there! Granted, if you're out finding your followers to take care of them, that's a good reason to be gone, but **you have to make yourself available to your followers if you're going to build trust.**

I may sound like a hypocrite here, but what I'm about to share with you is true and powerful. Each of these stories requires that you consider them in their proper context and purpose.

I was honored to serve as the Senior Enlisted Advisor to the Commander of the 92nd Air Refueling Wing in

Spokane Washington, 1994-1996. The wing had just been through the most terrible times possible about ten days before starting my tenure there. We had a shooting at the base hospital that left four dead and twenty-three wounded one Monday morning. Friday of the same week we had a B-52 crash, killing all four crew members. Things were pretty tough the two years my family and I were there. I never took a day of leave. As much as I tried to come up with ways to make life better for my Airmen there really wasn't a lot I could do…except maybe one thing…being there!

One morning, about zero-dark-thirty, which was, and still is, normal for me, I arrived at the White House (the wing headquarters) where I was greeted by the best First Sergeant I ever served with, Senior Master Sergeant Sonia Rejón, the First Sergeant for the largest group within the wing, the Maintenance Group. I have to admit, I was a bit anxious at seeing her sitting outside my office door. "Hola, Sonia," I greeted her.

"Buenos días, Chief," she replied.

"Cómo estás? What can I do for you, First Sergeant?" I asked.

"I need to talk with you, Chief," was her response. Uh oh! She sounded serious.

"Come into my office, please," I invited her. As we entered my office, she closed the door. *Man, this is going to be bad*, I'm thinking.

After sitting down and looking me in the eye, Sonia says, "Chief, I just want to tell you that my troops love you!" *WHAT?!* I didn't see THAT coming! "I don't understand, Sonia," I said, truly befuddled.

"You were out with them this morning at 0200 (2 AM)."

"Yes, I know. I do that as often as I can," I told her.

"Yes, we know. But no one else has ever done that. My troops love you for caring and if you ever need anything from us, you just ask. We've got your back!" She stood up without waiting for me to respond and left.

I sat there, bumfoozled, trying to remember what I'd done that morning with the Maintainers. All I'd done was hang out with them. I don't drink coffee, so I had a coke with them. I asked them how they, and their families, were. We joked around a little and they suggested a few things I could do to help improve their work and lives. I took note of those things with the intent of working them as soon as I could.

Sometimes, just being there is enough to inspire people to follow you. It shows you care. I've said it before…caring is the most important thing a leader does. There are as many ways to show you care as there are followers. Being there is one of them.

Life is one big paradox! What is, isn't. What isn't, is. I'm not being duplicitous, really. The importance of life's paradox is that both sides are true at the moment. But the moment changes everything. In one moment, this is the right thing to do. In another, THIS is the right thing to do. A real leader will reason out the moment and do what's right, now, right now! It's true! Being there is how you show your followers you care. And not being there is how you show your followers you care. See? Life is a paradox!

I hope you're young enough to remember watching The Andy Griffith Show. That show had so many lessons on leading that I could write a book on them. Maybe I will someday. I'd have to watch a lot of TV Land! Anyway, I remember an episode where Aunt Bea leaves for a few days to visit a relative and leaves Andy and Opie alone to fend for themselves. She prepares food and everything else they'll need to survive without her. Things don't go as well as expected while she's gone, but to make her feel better, just before she returns, Andy asks one of the neighbors, Clara, to come over and clean up the house and fix everything as if they'd been perfect stewards. Upon Aunt Bea's return, she walks into the house expecting to have to clean up and pick up after the guys, but instead finds the house perfectly clean. She's disappointed that they were fine without her.

Often, leaders expect their followers to need them to the point that they can't do without them. A real leader educates, trains, and empowers her followers to succeed even when she isn't there. In fact, that's the real test of her ability to lead. Will your followers do what they should do even without you leading and guiding them? If they do, you've done your job! You're a real leader!

There are moments when the best thing you can do is to be there. And there are moments when NOT being there is the best thing you can do. Ensure that you've done what you can to prepare your followers to lead themselves and then get out of the way! Whether it's at work or at home, for your followers or your children, you can only give them roots and wings. Often, it's easier to do their work for them, especially if you're an expert. But there's no learning in that. Teach them all you know and let them fly. They will! And they'll impress you. AND they'll follow you!

As Lao Tzu said, "A leader is best when people barely know he exists, when his work is done, his aim fulfilled, they will say: we did it ourselves."

So, building trust requires that you're there for your followers when you should be, when they need you to be. But it also requires that you back off and let them do what they know how to do. The best we can do for our followers is to give them roots and wings. And to know when to give them each of those. ***When your followers know that you're available when they need you, they'll trust you and follow you.***

If you make these eight lessons habits, they will lead you toward consistency. When we talked about sharpening the sword, I mentioned the term *moody*. Have you ever lived or worked with a moody person? How was life? I'm putting these questions in the past tense, hoping that you're not living in that environment now. Moody people have a huge negative impact on the environment. You may be asking, "What about people who are consistently moody?" I guess you can at least trust that they won't be positive. I stay away from those folks! Be consistent in all of the positive things, especially about what you've just read regarding building trust, and you'll soon be an Exceptionally Powerful Lieutenant.

One of the many things you'll struggle with even in just plain life, but especially as a leader will be keeping a balance between your work and your life. Both affect each other and as much as you'll try and separate them there will be days when it's near impossible.

Here's the good news…you CAN do it all! Here's the bad news…not all at once!

Whoever came up with the idea of multi-tasking has messed us up as a culture. We can't do that! There's plenty of research on the subject that proves that we can't do a lot of things at once. When we try to do so, we do most, if not all, of those things less efficiently. Sorry, but that's the truth!

Knowing that, then, how do we balance work and family? First, we have to define balance. We have this vision in our heads that balance is like the scales of justice. Two parts, evenly divided, fifty-fifty. Work and family balance is almost never fifty-fifty. That is, we don't invest fifty percent of our lives at work and fifty percent at home. Have you ever walked across a balance beam, maybe in gym class? Did you just walk across it, evenly? Or did you teeter this way and that way, until you reached the end? More the latter, right?

That's the way we balance work and family, we teeter a little this way, then that way, then this way again, depending on what's happening at the moment. The key, though, is staying focused on the moment, staying present and doing what should be done right now. It starts with knowing what's really important to us, our values, and doing that as best we can.

Here's an example. My family is way more important to me than my work. Sorry! I love you, teammates, but not that much! When my first granddaughter, Nieves, used to stay at our home, she'd come down the stairs from her upstairs bedroom and I would stop her at the bottom stair, hug her good morning, then carry her to my office, which is right next to the stairway. I'd have her sit on my lap for a couple of minutes and we'd just sit there, enjoying the mutual solitude. The choice for balance is in this…my computer, which is on the desk next to that chair, would carry on accepting emails and messages. Ding! You have mail! I had to choose whether or not to turn around and see what the message was or to sit, quietly, with my granddaughter, whom I say is more important. Man, but what if POTUS writes me asking me for my advice? I, surely, don't want to miss that email! Where did I say balance starts? With our values! I ALWAYS chose Nieves over those emails. And you know, we both felt better having done so.

Here's another example. The boss comes into your office and asks you to stay an extra hour to complete a few tasks. What do you do? Most of us would agree to stay. But is that what you SHOULD do? You want to balance your work life with your family life? Choose wisely! Maybe you should tell the boss that you have plans to take your spouse out for dinner in an hour, to celebrate your anniversary. Maybe the boss doesn't know that your daughter is performing her first ballet recital this evening. I'm convinced that most bosses would understand, them having been there, too, and would let you go. You have to choose! (Remember that keeping the boss informed will help you develop trust.)

Once you've made that right choice, focus on it. Don't be at your anniversary dinner looking at your iPhone. In fact, leave it in the car, if you have to carry it at all. The balance in your life will come from being present with your spouse for that short period of time. The work will still be there when you get back to the office, trust me.

Balance is a choice. Choose what's most important at the moment and do that. Actually, you will. Choose what you SHOULD choose, given your values. Sometimes, it will be work. But not most times. Balancing work and family is going to be dynamic. When you choose to be with family, be there, not half there, completely there. Same goes for work, but you probably already have that down, don't you?

Balance is a choice! And it becomes a skill the more you practice it. Like most skills, it may take some effort, but, man, is it ever worth it.

Developing and maintaining a work/life balance will lead you to developing consistency. Your followers will appreciate knowing (and trusting) that you don't go off the deep end constantly and haphazardly, that you're steady and in control of yourself. They'll trust you and follow you.

The final element for building trust— being principle centered—goes back to your

corps' core values. Principle centered people live those core values daily. That's not always easy to do. Life is simple but not necessarily easy. Leadership is the same way. So many leaders preach their corps' values yet don't live up to them. Your followers will see right through that. Trust—the glue that keeps you together and moves you in the right direction—will fade quickly. Know the principles that your corps expects you to embody and practice them daily in everything you do. The second-highest principle that all services embrace is integrity. (I'll soon tell you which is the highest.) Now, I'm not going to give you a lecture on integrity, you've already been through enough of them, but I will encourage you to read my book titled, *Beyond the Little Blue Book*, that provides possibly a different, for sure, an application-level, perspective on the topic. Practicing integrity will take discipline.

We all practice discipline on a daily basis in small ways that make us who we eventually are. As Socrates said, "We are what we repeatedly do." And what we repeatedly do is based on discipline of some sort. Discipline, by definition, is a controlled behavior. A behavior is what we do. We, no kidding, control EVERYTHING we do. The important question in developing and practicing the RIGHT discipline is WHY? (My friend, Simon Sinek, is smiling right now!)

The reason we do what we do is that it affects either us or them, positively or negatively. That's it!

Here are a couple of disciplines I practice daily, and why. I enjoy driving…fast. Deb won't let me drive the way I like to, though. I really enjoyed driving on the German autobahns, especially A-65, which had no speed limit. Man, get a BMW out there and fly! Love it! I can't, however, do that in Colorado Springs. For one, it's illegal, and for the other, Deb has a gadget on my cars that tells her when I've gone over 80 MPH. No kidding, there's an app for that! I see speed limits as more of a guide than a law. Except on base!

I served my country for 50 years. One of my most favorite groups of Warriors is the Cops. I still call them that, and they let me. Don't get me wrong, I love ALL Airmen, but I used to see the Cops every day. In fact, they, often, went by my office. Love those people! As I told you before, the word, *discipline*, comes from "disciple." A disciple is a follower. One who believes in something or someone. I believe in those men and women who put their lives on the line for us every day. They're the first line of defense in case we're threatened.

Anyway, as Deb and I drove on to USAFA yesterday, the first persons we saw at the gate, believe it or not, were Cops! I didn't have my ID out so Deb asks me if I know them. As I get up to the gate,

I stop, say hello to A1C P (can't' pronounce his name) and A1C Fleming. Fleming comes up to me and I tell him what Deb just asked me. He laughs and says, "We ALL know you, Chief!" He made me feel special.

As we drove on the base, I made sure I kept to the speed limit. Had to! I love those two Airmen, and all of their Peeps. I WILL NOT violate the law and put them in a situation where they'd have to admonish me! I will practice the discipline of maintaining the posted speeds because I believe in them! Now, I often argue with myself, you know that internal argument we have with invisible people, that the speed limits should be higher, but until they get changed, I'm not breaking them. By the way, I also use my turn signals!

We practice discipline all the time. We just don't think about it and we, surely, don't think about why, other than we may get punished if we don't. Think about why you do what you do, and that those behaviors affect others.

Every time I drive on base, I keep the speed limits. I can control that behavior. It may seem small, but I know my Brother and Sister Cops will appreciate it and me. I am a disciple of the Defenders of the Force! And, just as importantly, the principle of integrity, doing what's right.

There's a difference between discipline and punishment! If you're striving to be an Exceptionally Powerful Lieutenant, you should know the difference, and strive for discipline. Here's the difference.

We wonder why we have leadership problems in our culture, yet our perspective on developing followers is, mostly, based on forcing people to do what we want them to do. How's that working for you? It may have worked decades ago, but it doesn't work now.

Having served in the United States Air Force for fifty years, I often witnessed a method of "disciplining" that included making sure everyone paid for the indiscretions of a few, often, one. And, man, did that ever work! NOT! I remember having a Family Day scheduled for the upcoming Friday, which would give everyone a three-day weekend. Everyone was a disciple of the Wing Commander! He was THE MAN! (He happened to be male.) Until, however, comma, (Ever notice that nothing positive ever follows "however, comma"?) an Airman got caught DUI the Tuesday before. In his attempt to "discipline" the Airman, and the wing, the commander decided to take the Family Day away! WAIT! One person messed up and everyone has to pay?! That's PUNISHMENT! When this happens, and it still does, the intent is that everyone will understand that we're a team and that the actions of one affect everyone. We have to be wingmen and ensure everyone does what's right. You'd think we'd know better! Punishing everyone for the actions of one has never worked, and never will. Sorry, but that's the truth!

Now, I'm not advocating anarchy! What I AM advocating is that as a leader, YOU consider developing your followers into people who follow you for the right reasons and do the right things even when you're not around. That's discipline. And that's leading! There's a difference between discipline and punishment. Know the difference. Practice discipline! Lead effectively! That's being principle-centered.

Let me share the truth with you again. The highest principle is love. I know that you don't want to get into that because you've been taught that warriors are macho, and this is a touchy-feely subject. Remember my asking you about what's most important to you? Your response was family. Isn't family touchy-feely? It is for most of us. Even your corps' values are touchy-feely because they have to do with touching others in a positive way, and it feels so good when everyone in the corps lives up to them. Let me remind you that I define the *L* word as *unconditional giving*. **If you're principle centered, you'll give all you have, unconditionally.** You'll love your people and appreciate them for who they are, and they will do the same in return.

As I said before, it takes a combination of these behaviors to develop trust. The more of them you practice, the more trust you will have—and the more effective you will be. Now, let me give you the real lesson in trust. Remember when

I asked you to define the term and you came up with ways others display trust? Here's the truth, the bottom line. *The only way to develop trust is for you to be trustworthy.* *You* have to do these eight things you nodded your head about. *You* have to be willing to make the effort to be trustworthy. Others don't develop trust. *You* do! Do it now! Do it right! Make it last!

War Stories

Here's some advice from my protégé, Major Henry Sims, who served as an enlisted Marine and Airman before becoming an Air Force Officer.

Organize [Trust] and Equip - OT & E

As leaders in the U.S. Military, we place special emphasis on our teams by ensuring they are Organized, Trained, and Equipped to perform the roles the nation entrusts them to do. While training is one of the most important aspects to enable mission accomplishment, I also believe that building trust is the foundation of basic leadership. So, my spin on OT & E is incorporating trust into the equation. During a podcast with Chief (Retired) Bob Vásquez, we

BOB VÁSQUEZ

discussed the importance of trust. As a result, I recalled a 'war story' regarding an experience I had with a superior.

There I was, in Djibouti, Africa, leading a completely Joint Division consisting of members from every branch of service. It was an amazing team that conducted 24/7 system monitoring in the Network Control Center. I led the only 24/7 Division in the Directorate, which meant their equipment was utilized around-the-clock. One day, the time came to submit Unfunded Requests we call UFRs (yōō-furs). This presented opportunities for leaders to scour their respective areas for things they would like to purchase for their respective teams. The *only* thing my team wanted/needed was new chairs. The chairs they were currently using were horrendous. For months, they had taken the initiative to repurpose parts from other broken chairs to create ones that were functional. Sounds like a reasonable request, right? Considering other units were ordering things they wanted versus things they needed, I knew this request would not receive a second thought. I promised my team I would make it happen. This was my chance to earn *their* trust by ensuring they were Organized and Equipped.

Well, it wasn't so easy. The Deputy Director I will refer to as CDR Stout (Commander, Navy O-5) immediately questioned my request for new chairs.

244

After several email and phone exchanges, I attempted to convince him of my team's needs to no avail. Ultimately, he stated, "I am going to come over and see them for myself." From this perspective, he did not trust me (due to previous personality conflicts). To make matters worse, he came over and publicly questioned my request in front of my team. He clumsily moved chairs around, sat in them, and accidentally pulled a broken headrest off that had been temporarily secured by a team member.

I did not back down and neither did he. Regarding the broken headrest, he demanded, "Well give them the one from your chair." "I already have," I responded. After a public stalemate, we both went into my office to exchange frustrations. He expressed his dislike for my leadership style and at one point made a pointing motion in the direction of my face. Based on my immediate body language and my direction to end our conversation, he immediately realized he crossed the line and departed. There is more to the story but let's just say, I got new chairs for my team. Fight for your people, earn their trust and always ensure that they are Organized and Equipped to do their jobs.

You're going to find this hard to believe, but I can be arrogant when I choose to be. Yeah, yeah, I know you're thinking, "Say it ain't so, Chief!" I promised you the truth, and that's the truth, as much as it hurts you. Let me prove it to you.

The Fifteenth Air Force Band at March AFB, California, was my first assignment as a Chief Master Sergeant. I was a bit proud of myself. I'd survived some events in my life that others had said would be fatal to my career. I'd beaten the odds and attained the highest rank in the Air Force's enlisted structure. I was in the top one percent of the enlisted force! To put it in the vernacular of the day, I was a little "ate up" with myself.

In almost 20 years of service as a musician, I'd never had the opportunity to go home to play for my folks—all of those countless hundreds of people who lived in Deming, New Mexico, and whose last names ended in *uez*. That opportunity came shortly after I arrived at March. The band hadn't toured near my hometown, so my commander suggested that I take the big band there to show some homeboy-makes-good pride. Great idea! Since that area of the Southwest was my turf, I decided to do the advance work, just to ensure a successful tour. I traveled to several small towns in southern New Mexico. (Okay, all towns in southern New Mexico are small.) Anyway, Deming was my last stop before returning to March. I did my work and stayed an extra day to be with my family. I hadn't seen some of my pals in years, so I went out searching for them. I was amazed at how many of my high school friends were still in town. (High school, by the way, was a great experience for me—the best six years of my life. Hey, I was an overachiever!)

As I told you at the beginning of this story, I can be arrogant. Look, my buddies made me that way. They'd gone nowhere in the past 20 years. Richey and Eddie Sainz, Hector Ochoa, Richard Acosta, and Charlie Sera had all stayed in our dinky little hometown doing what their dads had done all of their lives, which is what their dads' dads had done before. I, on the other hand, had already seen the world. I'd played for all of the presidents of the United States alive at the time! I'd been overseas! Man, I'd been to Burma! I was worldly! How could I not be arrogant? See? My friends made me that way! I enjoyed visiting with my old chums, but we just weren't the same anymore. I was in a different league now. I went back to California to coordinate the tour that would show my folks how well I'd done.

We started in northern Arizona and worked our way to central New Mexico. It was winter, and the weather didn't seem to want to cooperate with us. Not only did we have to load in and out in the cold and wet, but we also had long stretches of nothing but desert between towns. We played at Eastern New Mexico University near Silver City one night and left the next morning for New Mexico State University in Las Cruces, where we would perform that evening. What a blustery day! Twenty-five of us traveled in an MCI bus, following an 18-wheel trailer carrying tons of musical and electronic equipment. As we traveled down this two-lane winding highway through the middle of nowhere, we noticed that the equipment truck, a few miles ahead of us, had pulled over to the side. There wasn't much of a shoulder, but the driver had gotten out of the way as best he could. Wouldn't you know that all four cars registered in New Mexico happened to drive by that truck on that day! It was a tough situation, raining and cold. The truck—and now the bus—hindered traffic. You see, those four guys had never seen a situation like this, so they had to keep coming around to see what was happening.

The truck was "hard broke," a term airplane mechanics use that means "that vehicle ain't going *no*where!" All this happened in the days before anyone had ever asked, "Can you hear me now?" We had no cell phones and no way (other than driving 100 miles) to get to a phone to call for help. Luckily, a local sheriff (yes, he had a gun) stopped and carried me and an assistant to his office to call for a wrecker.

Unfortunately, it now seemed that there were more than four cars registered in New

Mexico, and they were all broken down. Trucks with every local wrecker service I called were out and wouldn't be back for hours. I didn't have time to wait. The sheriff suggested calling one in Deming, the nearest town. (If you were paying attention, you'll remember that Deming is my hometown.)

I had no choice, so I called Sainz's Wrecker Service, the only one in town—Richard Sainz proprietor. Richey answered my call. (Again, if you were paying attention, you'll recognize the name of one of my old buddies.) "Hey, Bobby!" he said after I announced myself, "How are you?" Well, we talked about other things than my predicament, but we did get to that topic eventually. I needed his help, and I needed it quickly. As we finished our conversation, Richey said, "I'll be right there!" And he was!

I don't know how he got there so quickly. Maybe it had something to do with all of the state cops being family, so he didn't have to worry about speeding. On our way back to the truck and bus, I started thinking. Here I was, a top dog in my society—the Air Force. I commanded, literally, thousands of troops, had almost countless resources at hand, and had done things that I can't tell you about without having to shoot you—all in all a highly successful life. Yet I was stuck with my team in the middle of the desert, and the only person available to help me was a skinny, unpretentious guy whom I had recently looked down upon for not doing "more" with his life.

You'll recall all my allusions to humility. I was humbled to the bone. My friend Richey, a humble man himself, had come to my rescue and, in a sense, had rescued the entire United States Air Force by enabling us to accomplish our mission. That truck had major problems, but with the help of a few humble men, we finished the tour and returned to our home base safely.

Richey epitomized every aspect of building trust. He knew what he was doing, and he made sure I knew that he knew. He told me what he was going to do after grasping the severity of the problem. He never discussed payment; he just wanted to get us to where we needed to be. He was there when I needed him. He did his duty out of love and respect. He was, and still is, a warrior with whom I would go to war any day, anywhere. I trust him that much!

Starting Points

• **Am I trustworthy?** The key here, is the I! The YOU! Do you epitomize, better yet, embody, those eight CHARACTERistics you just read about?

How? It's almost easy to say that we do all of those things. It's another to prove it. As often as you can, consider whether or not you practiced those behaviors and how. Self-assessment is critical for your growth. And don't think that you'll accomplish all of those behaviors to their maximum every day. You're human. You'll fail at some of those behaviors some days. But keep striving. The more you strive, the closer you'll get.

• **Am I consistent?** I've admonished you before and I'm about to do it big time in the next habit. Find someone, an accountability partner, who you can trust to tell you the truth about whether or not you're being who you strive to be. And if/when you fail, keep improving. One percent at a time, that's the best way.

• **Do I share what I know with my followers and my bosses?** He who teaches learns twice. The best way to learn is to teach. The best way to receive is to give. Life is a paradox. As you share what you learn, remain open for different perspectives. You're going to believe what you think is right based on your knowledge and experience. Others, especially your followers, will have different knowledge and unique experiences that you can assimilate into your own thinking and behavior. You'll be a better person by accepting what others know. As Dr Stephen Covey taught me, one plus one equals three or more.

• **Do I live my corps' values?** We see them plastered all over the walls, our corps' values. We're trained to recite them and even recall their definitions. But do we know what they mean and what they look like? I'm not trying to sell you anything, in fact, much of the profits from my book, *Beyond the Little Blue Book*, go to the Air Force Aide Society's Airmen's Emergency Fund to help out your followers when in need, but the reason I wrote that book was to share what the Air Force Core Values look like through stories and perspectives. Don't just learn what your core and corps' values are, consider what they look like and practice them. Discuss them as often as you can with your followers, your peers, and your leaders. And then, assess whether or not, and how, you practice them.

Words of Wisdom

Few things help an individual more than to place responsibility upon him and to let him know that you trust him.
Booker T Washington

Most of us know how to say nothing, but few of us know when.
Unknown

Trust in what you love, and it will take you where you need to go.
Natalie Goldberg

We're like blind men on a corner—we have to learn to trust people, or we'll never cross the street.
George Foreman

We are at the center of a seamless web of mutual responsibility and collaboration.
Robert Haas

You can't have science with one scientist.
Alan Kay

The best proof of love is trust.
Joyce Brothers

We have to build trust in peacetime so that we can assume it in wartime.
Maj Gen Gary "Dutch" Dylewski, USAF, Retired

BOB VÁSQUEZ

HABIT 8

HANG ON TIGHT! FIND AN ENLISTED MENTOR

I'm sitting at my office desk when, all of a sudden, I hear a commotion in the offices across from mine. As I walk out to see what's the matter, I find a colonel yelling at the top of his voice at *my* lieutenant! This is unacceptable! I respectfully ask the colonel if he and I can speak in my office. He, of course, agrees. (You don't say no to the Chief.) In my office, I ask the colonel not to do what he just did. As diplomatically as I can, I tell him, "Sir, whatever My Lieutenant did, I'm sure he didn't do it intentionally. He's a good man, so you can talk with him in a civil manner, and he will give you all he has. Please show him the respect he's due." "You're right, Chief. I was upset and wasn't thinking. Thanks," the colonel replies as he walks out and talks with—not to—My Lieutenant, who takes care of him. As the colonel leaves the building, My Lieutenant comes into my office and thanks me for saving his butt. He didn't need to thank me. He was *MY* Lieutenant!

Remember the scenes in the movie *We Were Soldiers* in which the young soldier walks by SGM Basil Plumley, played by Sam Elliott? The young man says, "Good morning, Sergeant Major!" Plumley responds, "How the *#^& would you know what kind of morning this is?" Later, the same thing happens, and Plumley again says something unprintable. I think the movie director was trying to make a point. The Sergeant Major is depicted as tough—maybe even crude. I'm not sure Sergeants Major or other senior noncommissioned officers (SNCO) are quite that way anymore. Oh, they're just as tough all right, but maybe not as crusty.

In developing your leadership skills, consider this: Who will you lead? Enlisted folks! Who in the military world could better tell you about the people you'll lead than an enlisted person? This is not rocket surgery. Yes, that's right! An enlisted person will give you the best guidance. Now, wouldn't it be smart to find one who's "been there?" That would probably be a Senior NCO. Let me say, though, that plenty of middle-tier and junior NCOs can mentor you, especially in their areas of expertise, but they may not be as seasoned as a Senior NCO. If you're going to be an Exceptionally Powerful Lieutenant, find an enlisted mentor and hang on tight!

Here are a couple of other examples that Hollywood has blessed us with. How about the scene in the movie *Glory*, right after Colonel Robert Gould Shaw (Matthew Broderick) has flogged Private Trip (Denzel Washington) for "deserting?" Who does Shaw go to for advice? He seeks out Gravedigger (Morgan Freeman), who is obviously the most mature of his enlisted men. Colonel Shaw sees his value and seeks his counsel. In fact, Gravedigger is so good that Shaw makes him a Sergeant Major! Who knows enlisted folks? Enlisted folks! Find one fast!

Let's go back to *We Were Soldiers*. Recall, if you will, the part in which Lieutenant Colonel Hal Moore, played by Mel Gibson, is at his wit's end. He can't get out of a situation he's found himself in, and it seems that he and his troops are destined to reenact the fate of the famous (infamous?) 7th Cavalry (General George Custer's former unit). He looks at Sergeant Major Plumley, who's been with him the entire battle, and asks, "I wonder how Custer felt?" Plumley looks straight at him and says, "Sir, Custer was a weenie (he uses a different word). You ain't!" This immediately reenergizes Colonel Moore to reach inside and do what he has to do to save his troops.

Maybe the best example of this idea of holding on tight to a seasoned SNCO is the scene in *U-571* in which young Lieutenant Andrew Tyler (Matthew

McConaughey) becomes the submarine's skipper by default after the commander is killed. The Lieutenant isn't doing very well at leading his men. Over a cup of coffee, Chief Petty Officer Henry Klough (Harvey Keitel) asks the lieutenant for permission to speak freely. The skipper, of course, does so. The Chief admonishes him, saying, "The Commanding Officer is a mighty and terrible thing, a man to be feared and respected. All-knowing. All-powerful. The Skipper always knows what to do whether he does or not." Will you always know what to do? Probably not, but you'll have a better idea if you're close to someone who has much more experience than you do—who knows your followers and has led them before. Find that someone fast and hang on tight!

You'll need a Sergeant Major Plumley, a Gravedigger, or a Chief Klough in your military life if you're going to be a successful leader. You've probably been told by more experienced officers to find a trust-worthy Senior NCO and learn as much as you can from him or her. If you follow no other advice, do follow that counsel!

Do you remember that in Habit 1 I mentioned the two types of enlisted folks you'll find in your unit (those who need you to lead them and those whom you need to follow)?

When you come into my unit, I don't care whether you were the top graduate. I don't care if you won athletic awards at the Academy. I don't even care where you got your commission. What I care about is whether you can lead or follow. I know you can't lead because you haven't had the experience to do so yet. So, you'd better be able to follow. Now, here's the trick. If you're willing to follow me, I'll make you a great leader. I'll take you under my wing and help you excel. Recall my story at the beginning of this lesson. To get to *you*, they'll have to get through *me*. Senior enlisted professionals are a bit parochial in that they'll fight for what's theirs. If you're *MY* Lieutenant, I'm not going to let anyone, or anything, harm you! I'm responsible for your success.

Now, combine your education with my experience. Is that a formula for exceptional leadership? I think you'll agree that it is. Here's what will happen: all the time that you're following me, I'm making sure that everyone in the unit—and anywhere else—knows you're the leader. It's a circle. You see, I don't need to tell everyone that I'm a leader; they already know that. I'll make sure they know that you're the boss. And I'll make sure they know that if they mess with you, they mess with *me*! There are two people in this world that they don't want to mess with. I'm both of them! Okay, there are now *three*!

I hope that by reading this book, you've come to the conclusion that humility is a common thread that connects all of these habits. Exceptionally powerful leadership requires a whole bunch of humility; therefore, Exceptionally Powerful Lieutenants are humble. The central core value that the United States Air Force tries to live up to is Service Before Self. Humility grows from a person who strives to serve others before serving himself or herself. It also comes from realizing how little one knows, especially compared to the combination of experience, knowledge, and wisdom of others. Lastly, it grows out of understanding and valuing others. Humble yourself to an enlisted person who can show you the ropes and guide you to success. That doesn't mean you relinquish any of your responsibilities or authority. It means, in fact, that you strengthen them by sharing them.

Now, let me be very honest for a moment and tell you that not all enlisted folks will be good mentors. But you're no dummy. You now know how to listen, and you know what it takes to be an effective leader, assuming you put those other seven habits into action. So, you can find an enlisted mentor whose purpose is the same as yours—to serve. Together, you will do great things!

One of the most poignant ceremonies we perform in the Air Force occurs hundreds, often thousands, of times a day. The other services probably have a similar ceremony. It's quite simple, but it illustrates the critical connection between officers and enlisted folks. Let me paint the picture for you. As you know, the mission of the Air Force has to do with flying airplanes. Flying starts with taking off. The pilot, an officer, will jump into the cockpit (the old guys will crawl into it), put the key in the ignition, and crank up the plane. (Okay, it's not quite that way, but you might understand what I'm saying better if I use that example.) Anyway, the pilot's in the seat, ready to go. Just before he or she takes off the emergency brake to start down the runway, an Airman (formerly an NCO) stands tall near the nose of the aircraft and gives the pilot a thumbs-up, signifying that the airplane is in tip-top shape and will get that pilot, crew, and anything else in that plane to their destination safely. The Airman then comes to attention and salutes sharply. The pilot returns the salute, solidifying the bond between the officer and the enlisted person. It takes both.

Hang on tight to that, or those, enlisted professionals who are willing to guide you toward your goal of being an Exceptionally Powerful Lieutenant. There are plenty of them around. *But where do you find them?*

I'm an intuitive. I'm usually correct when I feel like I can trust someone, or not. I'm convinced that we all have some intuition. Use it to find an enlisted mentor or mentors. Keep in mind that everyone has specific expertise and experience. Don't expect one mentor to know everything. That's not possible. Consider having several mentors in specific fields. Some may be subject matter experts, and some may be character or leadership mentors. First, observe, as best you can, that prospective mentor's appearance and demeanor. Is it professional? That's what you're looking for. Recall what you learned in Habit 1 about making a first impression. You'll make up your mind about your followers based on that same criteria. And very quickly. Talk and listen as you decide whether or not you're a good "fit." And ask your peers, especially those who have been around longer than you. Most will offer their opinions and advice. Enlisted warriors are taught and expected to mentor their leaders. They don't all do it well, but you can learn something from everyone.

Once you do find one or more mentors, listen and learn as much as you can from them, and teach them, too. It's a team effort this thing we call serving our country. The least you'll get is a different perspective, which is key to being an effective leader.

Deb and I have been married for forty-five wonderful and fulfilling years! Yeah, go ahead and applaud. I just did! In all those years we've NEVER even CONSIDERED divorce! NEVER, EVER! Murder? That's another story....

As most people who live together for long periods of time do, we, sometimes, have a disagreement. Until I realize she's gonna be right. Hey! That's why we've been together so long! The other day we were having lunch at our kitchen table looking out at the forest our home is imbedded in when she asked me if what was on the ground was a squirrel or a rabbit. I couldn't see it, so she tried to describe it to me. I still couldn't see it, so she tried harder. Eventually, I suggested that I couldn't see it because it was behind a tree. She almost scolded me as she admonished me, saying it wasn't. It was visible...to HER! She got so frustrated that she got up from her seat and came over to SHOW me where the varmint was. As she did that, she realized that a tree was, indeed, in the way so she couldn't see it, either.

Whether we're leading others, or ourselves, for that matter, or just living with others, we usually assume that everyone has the same perspective on things. OUR perspective. But when we take the place of the person that we expect to agree with us, we may realize that their perspective is different. And it might even be valid.

I should stop reading the news and social media, but I probably won't. All I read lately is ONE perspective. It doesn't really matter whether or not it's MY perspective, it's ONE perspective, that of the writer/influencer, whatever term is correct. I'm okay with that, I know that I'll get that from that particular source. But there's no growth in that. What bothers me is that there's never a united perspective, one that acknowledges that there actually IS another way of seeing the same thing or event.

A current buzzword I hear often is "diversity." I'm all for diversity. As often as I hear someone use that word, though, I realize that they really mean ethnicity. I'm okay with that, too. Depending on the intended message. Diversity is about difference. We can all be from the same family, place, and time and still think differently and see things differently. We have different perspectives. Again, that's okay. In fact, as a leader, I commend you to do what Dr Stephen Covey used to admonish us to do, and that's celebrate the differences!

Maybe my old memory has rusted a bit, but I don't recall ever hearing and reading so much hate in the world. But, you know? I can change that for the folks I live and work with. I can open my eyes and my heart to another perspective. Maybe I don't have all the information or see things as someone is trying to explain to me. If you intend to be a leader, I suggest you do the same. Not only accept that there may be a different perspective but seek it. You may learn and grow from it. It may even make you a better leader!

By the way, it was a squirrel.

Let me tell you, if you can find a Chief to guide you, consider yourself blessed. Chiefs will tell you the truth, like it or not. And you'll need that to succeed. Never, ever, hire a yes-man/yes-woman, NEVER! You want another view than yours alone when you're making decisions. I believe it was General George S Patton who said, "If two of us are thinking the same thing, we don't need one of us." And know this, a Chief is a Chief for life! He or she will guide and advise you for as long as you can connect with him or her. Once a Chief, ALWAYS a Chief and she/he will mentor you.

Deb and I had the best seats at Clune Arena, where our USAF Academy Cadets play basketball. They're right at the edge of the tunnel where the players come on and off the court. Basketball's my game!

At this particular game, we're sitting in our great seats, watching the men warm up when this happens. Every team at USAFA has at least one commissioned officer who travels with the team and is their academics advisor throughout the season. Academics are incredibly tough at USAFA so it's a good thing to have scholars with the cadets to help them keep up with their studies. Anyway, Deb and I are watching the men warm up when I notice the full-bird colonel (O-6) who travels with the team, wearing his Mess Dress, coming toward us. This happens to be the same night of the Wing Annual Awards Banquet. I assume he's wearing his Mess Dress because he's going to the Awards Banquet. In my mind, I commend him for making the time to come see his team before the Awards Banquet. As he approaches Deb and me, I notice that the button that holds his Mess Dress coat closed is unbuttoned. As he gets within hearing distance of me, I call him over. "Sir, can you come over?" "What's up, Chief?" he asks. We've known each other for many years. Now, we're not buds or anything. I know who he is, and he knows who I am. "Sir, your coat button is unbuttoned," I tell him. He looks up at me and gives me the dirtiest look! An if-looks-could-kill sort of look. He buttons his button and goes out the tunnel.

I was bumfoozled by his response! I didn't expect him to make a big deal about me helping him out, but I, surely, didn't expect him to be ticked off at me. A little, "Thanks, Chief" would have been appropriate.

As he walks off, a security guard, a retired Master Sergeant, who the colonel walked past on his way toward where Deb and I are seated, comes over and commends me. "Thanks for doing that, Chief!" It would have been inappropriate to say anything, but I'm thinking, "Why didn't YOU stop and correct him? You're a retired Master Sergeant!"

Deb is livid! "Why did you do that?! That was embarrassing!" I've just recently realized that when Deb asks me a question, it's not really a question, it's a statement. I hadn't learned that yet, so I go about answering her question.

I take out my wallet and show her my ID card. I ask her, "What does this say my rank is?" She's now bumfoozled by what I'm doing, but she answers, "Chief Master Sergeant." I go on, "When does it expire?" "Indefinitely," Deb replies. "EXACTLY!" I retort. "I'm a Chief! I will ALWAYS be a Chief! My duty, AS a Chief is to take care of people. Officers, enlisted, civilians, families... that's what a Chief does. It's my duty to tell that colonel that he's out of regs. I'm his wingman. I didn't appreciate how he took it, but I have to have the courage to do what's right. It's what I teach these cadets. I can't tell them to do what I'm not willing to do. Besides that, if the Chiefs found out I let that go, they'd excommunicate me from the club!" *Man, I'm GOOD!* I'm thinking.

Deb gives me what we call in my culture, El Ojo, The Eye. I sit down and don't say a word the rest of the game. We won, by the way.

I count on others, including Deb, to help me be my best. Sometimes I don't know I'm failing. I'm going to assume you're like me. I'd rather someone correct me than let me do something stupid and realize later that it could have been avoided. Take courage and be a good wingman, too. It's your duty as a leader!

There's a picture that depicts what enlisted warriors do for their officers. It's called *Sergeant's Valor*. You can look at it, and even purchase a copy, at this site: https://www.donstivers.com/product-page/sergeant-s-valor. It's a Civil War Sergeant helping save an officer on the battlefield. It's very powerful. Here's the story that the painting depicts.

> On September 19, 1864, at Winchester, Virginia a remarkable act of courage and compassion took place. The charge was over and had been repulsed. Company after company of the 2nd U.S. Cavalry had smashed against Confederate breastworks to no avail. Orderly Sergeant Conrad Schmidt of Co. K had seen his commander go down in the melee, his right arm shattered by three pistol balls. Bloody, dazed, Captain Rodenbough staggered to his feet not fifteen yards from the enemy line when he saw his sergeant racing to his rescue. Overcoming fear is the definition of courage, and Schmidt's actions that

day earned him the nation's highest award, the Congressional Medal of Honor.

I always encourage enlisted warriors to consider that it's what we do. We take care of our leaders as well as our followers. Develop and maintain that bond. I guarantee you that it'll be worth it.

Your relationship with enlisted warriors, can reap a couple of very cool accolades, that of selection as an Honorary Chief or the Order of the Sword. Both are bestowed upon officers who support the enlisted corps above and beyond the norm. Chiefs groups at their respective bases have their own criteria by which they select Honorary Chiefs. The Order of the Sword is a more official award that is governed by specific criteria depending on the command. Both recognize individuals that enlisted warriors hold in high esteem and wish to honor.

Find an enlisted mentor whom you can trust to tell you the truth, even if it hurts. Learn as much as you can from that person and share your knowledge with others. Remember that trust is a two-way street that starts with you. Take care of that follower, and he or she will take care of you. Know this: enlisted folks *always* take care of their officers. If you treat them with respect, they'll return it in spades. If you mess with them, they'll take care of you. Enough said.

War Story

I was working at my desk at Ramstein AB, Germany, just before the Air Force threw me out. (I did *not* retire; that would require a voluntary action.) My teenaged daughter, Elyse, was sitting in a chair next to the door. As the Support Group Superintendent, I worked closely with several officers, one of whom turned out to be my favorite lieutenant in the whole world. (I'll just call him lieutenant so that I don't embarrass Lieutenant (now COLONEL) Eric Carrano.) As I worked, My Lieutenant came into my office several times asking for information, direction, money, and so forth. Every time he walked in, I stood up. My daughter watched this interaction for a while before asking me, "Dad, why do you stand up every time the lieutenant comes in here?"

"He's an officer, Baby, and I'm an enlisted person. That's what enlisted people do." (I could tell that she was a little confused.)

"Aren't you a big guy in the Air Force, Dad?" Elyse asked.

"Yeah," I said, wondering where she was going with this conversation.

"And isn't he a little guy?" my daughter inquired.

"Yeah, in a way." I answered.

"So why do you stand up for him? Shouldn't he stand up for you?" She asked, confused. Here's what I told her. (Hey, she asked for it, so I *had* to give her the sermon!)

"An enlisted person is supposed to stand up for an officer when the officer enters the enlisted person's presence. That's what the book says. I've been a Chief for about a dozen years now, and the book never stopped me from going beyond it, if you know what I mean. I make it a habit to stand for *everyone* who comes into my office. I still do so even though it's become more difficult as the knees age. It's a sign of respect, regardless of rank. The way I see it, we all may have different functions, but each of us has the same value. Some would say that enlisted folks stand for officers as a sign of subordination. That's crap! No one is subordinate to anyone else. God made us all equal. In fact, one of the charters of a professional warrior is to support and defend the Constitution of the United States, which states that we are created equal. Subordination creates negative energy. Humility creates positive energy. Subordination is maintained by an external source.

It's involuntary. Humility comes from inside of us. It's real power because we choose it! I've never subordinated myself to anyone, and I never will.

"I didn't stand *for* the lieutenant; I stood *with* him! Interestingly, he stood up when I went to him too. We don't stand *for* each other; we stand *with* each other. The combination of my power and his creates a third, much greater, power that will break apart without one of us. I know that my 'place' is to stand *for* him, but my strength is in standing *with* him. I hope that this explanation gives you a sense of my point. A powerful leader is humble. He or she realizes that power lies in helping and supporting others. The more you expand your support, the stronger your leadership power will be because the more people you give to, the more people will give to you. Humility is key to powerful leadership. End of sermon!" She understood.

I hope that I made my point with my daughter and with you. We all have to work together to accomplish anything worthwhile. Don't make the mistakes others have—learn from them. The more seasoned an enlisted person is, the more mistakes he or she has made that you can learn from. The enlisted force *is* the backbone of every service. I say that with all humility. That backbone will support you and make you the head if you do what you should. You'll surely find that the bond you develop with that enlisted mentor will last forever. Years later, several promotions later, you'll need that expertise, and you'll make a call or send an e-mail. You'll find that the relationship hasn't changed from the time when you served together. It's almost amazing. We are all connected. Strengthen those connections. Hang on tight!

Starting Points

• **Is my ego strong enough to follow an enlisted mentor?** As I said in Habit 7, on building trust, we often confuse humility with ego. There's a huge difference between a strong ego and a big ego. People who have a big ego go around telling others how great they are. People who have a strong ego don't have to tell anyone anything.... Humility is based on purpose. If you're sincere purpose is to grow, to develop your leading abilities, then you won't be able to help but to be humble. Humble people realize that they don't know everything and strive to learn more. And the best resource for your growth as a leader is your followers.

• **Am I willing to trust my enlisted mentor to tell me the truth, even if it hurts?** Most "leaders" I've worked with or observed, although they professed differently, hired people just like them. They often talked about wanting their followers to be authentic. Yeah, that's a current buzz word. That's not true. Those "leaders" didn't want their followers to be authentic, they wanted them to be THEM! I think it's a natural inclination. DON'T DO THAT! As you search for an enlisted mentor look for someone who will be willing to help you grow and empower yourself to be your best self. Growth requires DIScomfort. Your best mentor will be respectfully honest with you. Seek someone who asks more than tells. Recall my story about my boss being upset because the band wasn't playing to their potential. All I did was ask questions to guide him toward the right answers.

• **Am I being honest with my enlisted mentor?** A strong relationship is always based on mutual trust. The more everyone in the relationship practices those tenets, the more that trust will exist. You and your mentor have to be trustworthy for either of you to empower each other. And it will always begin with listening empathically.

• **Do I expect my enlisted people to stand *for* me or *with* me?** They have to stand for you.

Whether or not they stand with you will be up to how you treat them. Practice all of these habits and they will stand with you, follow you, and even be willing to die for you.

Words of Wisdom

If I have been able to see farther than others, it was because I stood on the shoulders of giants.
Sir Isaac Newton

The block of granite, which was an obstacle in the path of the weak, becomes a stepping-stone in the path of the strong.
Thomas Carlyle

It is necessary for us to learn from others' mistakes. You will not live long enough to make them all yourself.
Adm Hyman Rickover

Change is inevitable; growth is intentional.
Glenda Cloud

*A great teacher never strives to explain his vision.
He simply invites you to stand beside him and see
for yourself.*
Bobby Ray Inman

He who is afraid to ask is ashamed of learning.
Danish proverb

*We must all hang together, or assuredly we shall all
hang separately.*
Benjamin Franklin

*Great leaders don't set out to be a leader…. They
set out to make a difference. It's never about the role
– always about the goal.*
Lisa Haisha

FINAL THOUGHTS

I've tried to give you a view from the eyes and hearts of the enlisted folks you'll soon be leading. Enlisted people think differently than officers—and rightly so. I may have said it before, but here it is again: enlisted personnel and officers have different functions but the same value. You will die without your enlisted folks, or you will have great military careers with them. It's your choice. Make the right one. As you do that, think about the eight habits I've commended to you. Develop and employ them so that the folks you lead will think and say, "I'll go to war with MY Exceptionally Powerful Lieutenant any day, anywhere."

As I said in the preface, these thoughts are based on more than 50 years' experience in dealing with lieutenants and cadets of every sort. My hope is that you'll take these lessons and really turn them into habits. The more you do, the more powerful you will become. In addition to these eight habits, you'll need four assets to become truly powerful. I didn't include them as habits because they're a little different. You can develop habits. I'm not sure you can develop what I'm about to share with you. I think that you discover them. I can't give them to you; they'll come from deep inside you. They are *desire, capacity, will,* and *action*.

In Habit 1, I told you that you can live that first-day scenario, but you have to make it so. You have to do the work. In Habit 8, I mentioned that humility is powerful because it's a choice. If you're going to be an Exceptionally Powerful Lieutenant, you must begin your quest with the *desire* to be one. That desire will grow out of your purpose. Remember that I asked you whom you will lead as a lieutenant? I asked you several times because it's a critical question. The answer is still Enlisted Warriors. **If your purpose is to take care of your charges, your enlisted people, the desire to be your very best will surface.** You can be an Exceptionally Powerful Lieutenant if you desire it. You'll need three other assets though.

Wherever you go, the environment will be different than the one you just left. Thank goodness, ha? What you do and how you do it may be totally different. The one thing that won't change is you! Let me repeat myself. I don't know what you're going to do, but I know who you're going to be. A lieutenant! An Exceptionally Powerful Lieutenant, I hope! You have the *capacity* to do whatever you desire. Habits 4, 5, and 6 will help you hone that capacity. You may encounter obstacles that could hinder you, including people who will work at keeping you from fulfilling your dreams. Don't let them stop you. Find that NCO or another mentor, regain your strength, and express your power. It's within you! *You can be an Exceptionally Powerful Lieutenant if you desire. You have the capacity!*

You may have the desire and the capacity to excel, but you're going to have to find the will to do so. You'll have to find it in your heart. Some of the habits, particularly 2 and 3, may take a great deal of willpower to attain. If you've already developed bad habits, you'll have to work twice as hard to replace them with good ones. But you can do it! At times your progress will stall. Don't give up. *You can be an Exceptionally Powerful Lieutenant if you desire. You have the capacity! Find the will!*

Finally, none of this will have any effect whatsoever unless you *do something*! Indian yogi and guru Paramahansa Yogananda said that "a wish is desire without energy." You can sit back on your butt, wishing all day long—but nothing will change. Or you can get off your "buts" and make great things happen. "But I can't!" "But I don't know how!" "But what if I fail?" If you fail, learn from it. As John C Maxwell says, "Fail forward!" Progress requires *action*. *You can be an Exceptionally Powerful Lieutenant if you desire. You have the capacity! Find the will! Do it now!*

I wish you an exceptionally successful life as a lieutenant and beyond. I hope that you will pass on what you learn to others so that our force will continue to grow. The profession you've volunteered to pursue is the most honorable of all professions. You'll never be fully compensated for the sacrifices you and your family will make in your pursuit to serve, but in the end, your legacy will be one that very few people even dream of.

Know this: You will be tested! It's inevitable. Your followers will test you. Your leaders will test you. Your peers will test you. Do these eight things you've just read about daily until they become habitual, and you'll pass the test. You'll never know when you'll be tested. You have to be ready.

Today's one of those days that I'm totally bumfoozled! My body is telling me something different than my brain. My body is telling me how I feel. My brain is telling me how I should think. We just switched our clocks forward an hour. Who made that up? You mean we just told the sun what time to rise? Evidently. It's amazing how much power we have! Or so it seems when we're governed by the clock.

The clock is a perspective that leaders use way too often in measuring success. How long have you been a leader? Does that matter? Probably not, but that's a measure of a leader's success, isn't it? Time measures efficiency, not effectiveness. Just because you've been a "leader" for a long time doesn't make you a good one, or one at all, for that matter.

The term the Ancient Greeks used for measuring time by the clock was called Chronos. Chronos is quantitative, it's easy to measure, as long as you have a clock or a sundial. It's sequential. One event follows another. Leading isn't that way, have you noticed? It's confusing when you expect someone, a follower, to do things a certain way, and she doesn't. And you're bumfoozled by why she didn't do it as you told her to. Maybe there was a better way? Maybe it wasn't the right time?

The other term the Greeks used for time is Kairos. Kairos means "the right, critical, or opportune moment." What time do you start leading? When you get to the job or the office? Or when it's the opportune moment? Do you lead only when you're on the clock, on duty, or when someone needs your help or support?

My favorite word in the English language is "serendipity." Serendipity means "the occurrence and development of events by chance in a happy or beneficial way." I'm constantly aware of serendipitous moments, especially when opportune moments to lead appear.

Leading effectively can't be based on a Chronos perspective. It has to be based on a Kairos perspective. When will that opportune moment appear in front of you? You won't know, but it will happen. Will you step up and lead? If you do, it will be beneficial to you and the person you lead, and it may make both of you happy.

I hope there's a lot more leading going on than we think. I hope you lead more often than you even know. I hope that you take that opportunity to support someone when THEY need it, when a Kairos moment serendipitously arises. Don't worry about when. Do it now. Right now's the moment. It's time to lead!

Here's the ultimate goal of leading, I believe. The thought was inspired by what one of my favorite authors, Max De Pree, wrote in his book titled, *Leading Without Power*. In it he refers to a term that I've used in my job resumé cover letters for years. I'd never seen it before, so I thought I'd created it. Evidently, great minds DO think alike. The term is Realized Potential.

My sense is that Realized Potential is the goal of Effective Leaders, of Exceptionally Powerful Lieutenants. What does that mean? I'm about to share my perspective with you. The full term that De Pree uses is "a place of realized potential." He says that that's THE PLACE we all want to be, and leaders are charged with creating it. Again, what does that look like? Here's my vision.

I looked up the word, "place," online. One definition was "a portion of space available or designated for or being used by someone." You've, surely, read a book or two in which the author refers to people's need for "space" to do this or that. There are plenty of moments at home and at the shop when I could use a little "space." A moment or two, usually to reflect, and maybe even learn, from something that just happened. I'm convinced that an Effective Leader creates that for her followers.

When we think of "space" we usually think of a physical place...here, over there, out there.... But space is also a moment in our lives, as I just mentioned, for self-assessment, for re-vectoring, where we feel safe and secure in being who we are so that we can do what we should do. If you've ever heard me speak, you've heard me say that "There's no place like this place anywhere near this place, so this must be the place!" I borrow that from Dr John C Maxwell. That statement DOES refer to a physical place, but there's also a spiritual/emotional space that Effective Leaders should create for their followers. That's what I aim to do... help others find that space within them. That's THE PLACE!

To realize means "to cause (something desired or anticipated) to happen." REALIZED means that it already did. Imagine you and I meeting up in December, or any time in between, to share whether or not we've realized what we resolved to do this year! By the way, I would find it an honor to meet up with you! Maybe one of your revolutionary resolutions was to show your followers that you really care. I suggest that creating a place, space, if you will, for your followers to think about who they are and what their vision is, would be a way to show them you care. So, this is how you've done it. This is how you've realized that vision. And you do it daily. That's effectiveness. That's leading! That's leading effectively! I'm so proud of you!

The best way to start leading effectively is to discover and empower your followers' potential. Each of your followers has it. But what is it? By definition, it's "having or showing the capacity to become or develop into something in the future." You knew that. I know. But do you know each of your followers' potential? I served my country for 50 years. I still remember my recruiter telling me not to volunteer for anything in Basic Training. He told me to just keep my mouth shut and do what I was told to do. Well, it wasn't in my nature to do that. But I graduated! You'd probably be amazed at what your followers are capable of doing. If you can find out what some of those things are and provide them resources to hone those skills, they will follow you anywhere! And isn't that a leader's goal, for his followers to actually follow? Find out what they can, and want, to do. Get them the resources. Get out of their way. Reap the benefits! Create a Place of Realized Potential for them!

Let me leave you with a thought from American author and clergyman Edward Everett Hale; it encapsulates all I said in the previous pages:

I am only one,
but I am one.
I cannot do everything,
but I can do something.
And because I cannot do everything,
I will not refuse to do the something that I can do.
What I can do, I should do.
And what I should do, by the Grace of God, I will do.

Be GREAT! You ARE! HEIRPOWER!

Chief bob vásquez!

Other books by bob vásquez available at amazon.com and Apple Books...

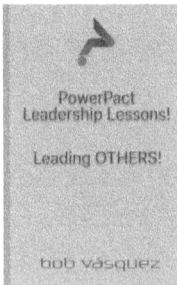

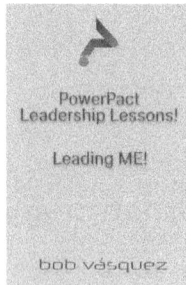

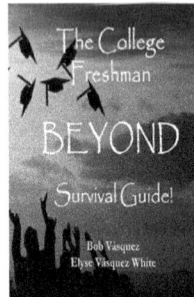

www.ingramcontent.com/pod-product-compliance
Lightning Source LLC
Chambersburg PA
CBHW030004290326
41934CB00005B/219